FLORA OF TROPICAL EAST AFRICA

HYMENOPHYLLACEAE

Henk Beentje[1]

Epiphytic, lithophytic or terrestrial ferns of medium to small size; rhizome usually creeping, seldom erect, bearing hairs. Stipe terete, not articulated, often with a single leaf trace; lamina simple to pinnately compound to irregularly divided; segments single-veined; lamina usually a single cell thick, lacking stomata; veins free or joining to form a submarginal vein; false veins, unconnected with the real ones, sometimes present. Sori terminal on veins, solitary at apex of terminal segments or marginal on simple to pinnatifid fronds; involucre cup-shaped to deeply 2-cleft; sporangia short-stalked to (sub-)sessile, maturing basipetally and carried on receptacles terminating a vein, these short, capitate or clavate to long and exserted; annulus oblique, not interrupted; dehiscence irregular. Spores globose-trilete, tetrahedral, containing chloroplasts, usually short-lived.

600 species in either two genera (*Trichomanes* and *Hymenophyllum*) or into various more 'natural' genera.
The family is most diverse in the mossy montane forests of the tropics and south-temperate areas.

Here I follow the recent treatment by A. Ebihara, J.-Y. Dubuisson, K. Iwatsuki, S. Hennequin & M. Ito in Blumea 51, 2: 221–280 (2006), which is based on a world-wide revision based on molecular data, and which is an approach which tries to recognise natural (monophyletic) as well as morphologically recognizable groups. This supersedes the slightly older work by Iwatsuki in Families & genera of vascular plants 1: 157–163 (1990), taken up in recent Floras such as Flora of Australia 48 (1998) and by Roux in Conspectus of southern African Pteridophyta (2001).

1. Lamina, stipe and rhizome with stalked stellate hairs; involucre bivalvate . . . . 1. **Hymenophyllum**
   Lamina and stipe glabrous or rarely with a few simple hairs 2
2. Rhizomes glabrous or nearly so; involucre bivalvate . . . . 1. **Hymenophyllum**
   Rhizomes covered with dense simple or simply branched hairs; involucre conical or cylindric 3
3. Rhizome short-creeping to erect . . . . 6. **Abrodictyum**
   Rhizome long-creeping 4
4. Rhizome thick, 1–5 mm in diameter, with abundant rootlets and with dense dark articulate hairs to 3 mm long . . . . 5. **Vandenboschia**
   Rhizome filiform, < 1 mm in diameter; rootlets sparse, hairs up to 1(–1.5) mm long 5

[1] Royal Botanic Gardens Kew.
I am grateful to Koos Roux for his excellent Conspectus of southern African Pteridophyta, from which I have used the generic descriptions.
Both Dr Verdcourt and Dr Ebihara read through my manuscript for errors, and I thank them.

5. Continuous false veinlets parallel to true veins; venation
   catadromous (first set of veins in each segment of the
   frond originates from the basiscopic side of the midrib)   **2. Didymoglossum**
   False veinlets absent; venation anadromous (first set of
   veins in each segment of the frond originates from the
   acroscopic side of the midrib) 6
6. Rootlets absent; rootlet-like shoots present .... **3. Crepidomanes**
   Rootlets present .... **4. Polyphlebium**

(I have keyed out the last taxon under *Crepidomanes* as well, as this is not an easy
character to see)

1. **HYMENOPHYLLUM**

Sm. in Mem. Acad. Sci. Turin 5: 418, t. 9, fig. 8 (1793); Ebihara et al. in Blumea 51,
2: 226–234 (2006)

Sphaerocionium C.Presl.

Rhizome creeping, irregularly branched, with few scattered roots. Fronds spaced;
stipe terete, often winged in upper part; lamina pinnately compound, glabrous or
with few small trichomes; ultimate segments 1-veined, venation free, anadromous,
without false veinlets. Sori at apex of ultimate segments; involucre deeply cleft;
receptacle included or exserted.

± 250 species throughout the tropics and temperate zone.

1. Lamina and stipe with stalked stellate hairs 2
   Lamina and stipe glabrous or rarely with a few simple hairs 4
2. Lamina stellate-hairy over whole surface; rachis winged ...   *1. H. splendidum*
   Frond surface glabrous, hairs (stellate or simple) present
   on margin and costa 3
3. Rachis not winged, the pinnae slightly decurrent .... *2. H. capillare*
   Rachis winged, the wings continuous .... *3. H. hirsutum*
4. Ultimate segments serrate or spinulose-denticulate 5
   Ultimate segments with entire margin (may have few dark
   trichomes) 7
5. Lamina 40–60 mm wide; sori 2–3 mm long .... *4. H. triangulare*
   Lamina 10–35 mm wide; sori 1.1–2 mm long 6
6. Pinnae mostly acroscopically divided; involucral lobes ±
   entire .... *5. H. peltatum*
   Pinnae divided on both sides; involucral lobes dentate ... *6. H. tunbrigense*
7. Lamina 2–3-pinnatifid, much longer than wide 8
   Lamina palmately flabellate or irregularly lobed, about as
   long as wide 9
8. Pinnae of mature fronds mostly with fewer than 15 lobes   *7. H. capense*
   Pinnae of mature fronds mostly with more than 20 lobes   *8. H. kuhnii*
9. Lamina margin glabrous; sori valves erose-dentate ...... *9. H. sibthorpioides*
   Lamina margin with a few dark trichomes to 0.3 mm; sori
   valves entire .... *10. H. sp. A*

1. **Hymenophyllum splendidum** *Bosch* in Nederl. Kruidk. Arch. 5: 192 (1863);
Benl, Pter. Fernando Po: 8 (1980). Type: Bioko [Fernando Po], *Mann* s.n. (K!, holo.;
BM, L, iso.)

Epiphyte or lithophyte; rhizome creeping, branched, filiform, densely hairy with red-brown long-armed stellate hairs to 1 mm long. Fronds spaced 1–6 cm apart, pendulous; stipe 2–10 cm long, winged in upper part, stellate-hairy; lamina dark green turning brown, linear-lanceolate in outline, 8–45 × 1.7–6 cm, 2-pinnatifid; pinnae in 8–24 pairs, rhomboid-oblong, to 3 × 1.2 cm; ultimate segments 6–15, more numerous along acroscopic margin, linear, to 2.5 mm wide, rounded at apex, denticulate with the projections bearing hairs; rachis winged throughout; lamina hairy throughout with yellowish stalked-stellate trichomes. Sori apical on terminal segments in distal half of lamina, 1–1.6 mm in diameter, indusial valves ciliate with branched hairs; receptacle included. Fig. 1: 1–4; 9–12, p. 4.

UGANDA. Ruwenzori, above Kilembe water supply intake, Feb. 1967, *Lock* 69/12!; Kigezi District: Ishasha R., 7 km SW of Kirima, Sep. 1969, *Faden et al.* 69/1225!
KENYA. Embu District: Irangi Forest Station, Apr. 1972, *Faden et al.* 72/186!; Mt Kenya, Castle Forest Station, Oct. 1979, *Gilbert* 5782!; Teita District: Kasigau, June 1969, *Napper et al.* 2175!
TANZANIA. Kilimanjaro, Weru-Weru Gorge, Dec. 1996, *Hemp* 1373!; Kilosa District: Ukaguru Mts, Mnyera Mt, June 1978, *Thulin & Mhoro* 2831!; Iringa District: Udzungwa Mts, Mwanihana Forest Reserve, Oct. 1984, *D.W. Thomas* 3867!
DISTR. **U** 2; **K** 4, 7; **T** 2, 3, 6, 7; Bioko, Cameroon and east to Kenya, south to Malawi, Mozambique and Zimbabwe
HAB. Moist forest, where often near water; low epiphyte or on moist rock faces; may be mat-forming; 1200–2400(–3000) m
USES. None recorded
CONSERVATION NOTES. Widespread; least concern (LC)

SYN. *H. ciliatum* Sw. var. *splendidum* (Bosch) C.Chr., Index Fil.: 368 (1906); Taton in Bull. Soc. Roy. Bot. Belg. 78: 16 (1946)
Sphaerocionium splendidum (Bosch) Copel. in Philipp. J. Sci. 67: 31 (1938); Burrows, S. Afr. Ferns: 100, fig. 101, map (1990); J.P. Roux, Conspect. southern Afr. Pteridoph.: 44 (2001)

2. **Hymenophyllum capillare** *Desv.* in Mém. Soc. Linn. Paris 6, 2: 333 (1827), based on *T. hirsutum* Thouars in Fl. Tristan Ac.: 34 (1804), *non* L.; Taton in Bull. Soc. Roy. Bot. Belg. 78: 19 (1946); Alston, Ferns W.T.A.: 32 (1959); Tardieu in Fl. Cam. Pterid.: 75 (1964); Schelpe in F.Z. Pterid.: 80 (1970); Benl, Pter. Fernando Po: 7 (1980); Burrows, S. Afr. Ferns: 99, fig. map (1990). Type: Tristan da Cunha [Tristan d'Acugna], without collector or number (not found) (see Note)

Pendulous epiphyte or hanging from rocks; rhizome long-creeping, filiform, glabrous or with long-armed stellate hairs. Fronds 1–5 cm apart, pendulous; stipe 1–6 cm long, terete, filiform, with partly deciduous branched stellate hairs; lamina linear in outline, to 6–70 × 2–3(–4) cm, 2–3-pinnatifid; rachis terete almost to apex, not winged, with stellate hairs; pinnae up to 30 on each side of the rachis, to 2 × 1 cm, the lowermost smaller and remote; pinnules up to 9, more developed on acroscopic side, linear, entire or lobed, ultimate segments to 9 × 1.5 mm, rounded to slightly emarginate, margin entire, with stalked stellate hairs on margin and veins. Sori at the lobe apices in distal part of pinnae, obconic with rounded valves, 1–1.6 mm in diameter, pubescent with stellate hairs; receptacle included. Fig. 1: 5–8, p. 4.

UGANDA. Ruwenzori, Nyinabitaba–Kichuchu, Jul. 1951, *Osmaston* 3864! & Nyinabitaba Hut, Dec. 1950, *G. Wood* 225!; Ankole District: Kasyoha–Kitomi Forest Reserve, near Nzozia R., Jun. 1994, *Poulsen et al.* 624!
KENYA. Meru District: Nyambeni Hills, Kirima Peak, Oct. 1960, *Verdcourt* 2947!; Kericho District: SW Mau Forest, 16 km SSE of Kericho, Kiptiget R., June 1972, *Faden & Grumbley* 72/335!; Teita District: Kasigau, main peak, Nov. 1994, *Luke* 4214!
TANZANIA. Arusha District: Mt Meru, Jekukuma R., Mar. 1966, *Greenway & Kanuri* 12464!; Morogoro District: N Uluguru Forest Reserve, Palu, Dec. 1993, *Kisena* 985!; Rungwe District: Rungwe, Aug. 1911, *Stolz* 873!
TANZANIA. Kilimanjaro, Old Moshi, above Kidia, Dec. 1995, *Hemp* 941!; Lushoto District: E Usambara, Mazumbai, Nov. 1974, *Balslev* 303!; Iringa District: Udzungwa Mountain National Park, Luhomero Mt, Oct. 2000, *Luke et al.* 6982!

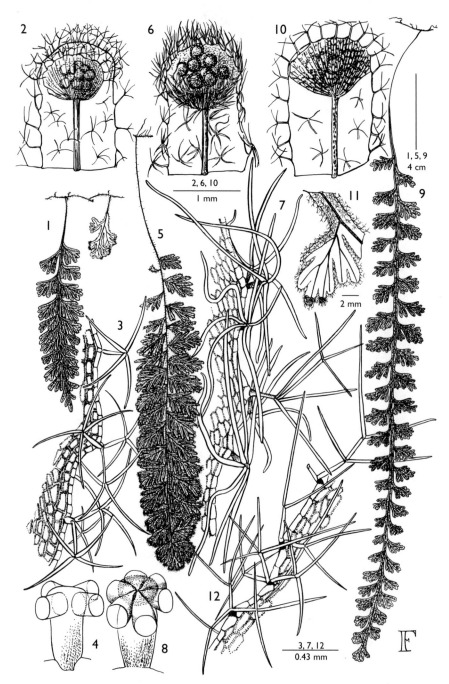

FIG. 1. *HYMENOPHYLLUM SPLENDIDUM* — **1**, habit, × ¹/₂; **2**, pinnule apex with sorus, × 18; **3**, lamina margin, much enlarged; **4**, insertion of trichome branches, diagrammatic. *HYMENOPHYLLUM CAPILLARE* — **5**, habit, × ¹/₂; **6**, pinnule apex with sorus, × 18; **7**, lamina margin, much enlarged; **8**, insertion of trichome branches, diagrammatic. *HYMENOPHYLLUM SPLENDIDUM* — **9**, habit, × ¹/₂; **10**, pinnule apex with sorus, × 18; **11**, pinnule, × 2; **12**, lamina margin, much enlarged. 1–4 from *Greenway* 6545A; 5–8 from *Gilbert* 5781; 9–12 from *Faden et al.* 70/591. Drawn by Monika Shaffer-Fehre.

DISTR. **U** 2; **K** 3–5, 7; **T** 2, 3, 6, 7; widespread in tropical Africa from Ghana to South Africa; Mascareignes, Madagascar

HAB. On treetrunks, less often on mossy branches and moist cliff faces, within moist forest, montane evergreen forest and tree heath; may be mat-forming and often locally common; (1000–)1650–3150 m

USES. None recorded

CONSERVATION NOTES. Widespread; least concern (LC)

SYN. *Hymenophyllum holotrichum* Peter, F.D.-O.A.: 16 & Descr.: 1, t. 1.3, 4 (1929). Type: Tanzania, Kilimanjaro, near Moshi, *Peter* 1268 (B!, lecto.)

    H. anisopterum Peter, F.D.-O.A.: 15 & Descr.: 1, t. 1.2 (1929). Type: Tanzania, Kilimanjaro above Bismarck Hill, *Peter* 890b (B, holo.)

    Sphaerocionium capillare (Desv.) Copel. in Philipp. J. Sci. 67: 33 (1938); J.P. Roux, Conspect. southern Afr. Pteridoph.: 44 (2001)

    Sphaerocionium capillare (Desv.) Copel. var. *alternialatum* Pic.Serm. in Webbia 23: 196 (1968). Type: Kenya, Mt Kenya, Kathitha R., *Schelpe* 2592 (Herb. Pic.Serm., holo.; BM, BR!, P!, iso.)

    Hymenophyllum capillare Desv. var. *alternialatum* (Pic.Serm.) Faden in U.K.W.F.: 28 (1974), **syn. nov.**

NOTE. Roux in his Conspectus (under *Sphaerocionium capillare*) has as type Reunion [Ile Bourbon], *J.M.C. Richard* s.n. (P!, holo.) but this specimen is not mentioned in the protologue and so cannot be the type – I have not seen a designaton as lectotype.

    Pichi Sermolli described his var. *alternialatum* based on a specimen with long narrow fronds, with the pinnae set apart and long-decurrent; intermediates with the typical taxon occur for both spacing and extent of decurrence and I have included the variety in the synonymy.

    Several specimens at major herbaria such as Berlin and Paris have been named as a forma or a variety *major* Rosenst. or *majus* Rosenst. – some of these have been identified as such by Rosenstock himself, some by Bonaparte. I do not believe the name has been published. There seems to be no difference from other material in this species.

    Var. *alternialatum* was distinguished by having the pinnules slightly decurrent on the rachis. I do not believe this warrants varietal status.

3. **Hymenophyllum hirsutum** (*L.*) *Sw.* in J. Bot. (Schrader) 1800, 2: 99 (1801); Benl, Pter. Fernando Po: 9 (1980). Type: based on Plumier, Traite Foug. Amer.: t. 50/B (1705), iconotype

Epiphyte or lithophyte; rhizome creeping, branched, filiform, with dense pale long-armed stellate hairs to 1 mm long. Fronds spaced 1.5–7 cm apart, arching, eventually pendulous; stipe 1.5–14 cm long, winged in upper part, with simple or branched reddish hairs, glabrescent; lamina dark green, deltate to oblong-ovate in outline, 10–24(–40) × 2–8 cm, 2–3-pinnatifid; rachis and costae winged; pinnae in 11–22 pairs, ovate to linear in outline, 1–3 cm long, decurrent, the lower ones shorter; ultimate segments linear, 4–6 mm long, 1–1.5 mm wide, obtuse; rachis, costa and veins with stalked stellate trichomes; lamina margins with trichomes or trichome remnants; lamina surface glabrous. Sori 1–2 per segment, terminal, on upper part of lamina, 1–1.5 mm in diameter; involucre cleft to base or nearly so, ciliate with simple and branched hairs; receptacle included.

UGANDA. Toro District: 5 km W of Kilembe, June 1970, *Lye & Katende* 5532!; Kigezi District: Ishasha R., 7 km SW of Kirima, Sep. 1969, *Lye et al.* 4191!

TANZANIA. Pare District: S Pare Mts, Mtonto, July 1942, *Greenway* 6545a!; Kilosa District: Ukaguru (Itumba), Mar. 1905, *North Wood* s.n.; Morogoro District: Nguru Mts, above Mhonda mission, Jan. 1987, *Schippers* 1692!

DISTR. **U** 2, 4; **T** 2, 3, 6, 7; Liberia to C.A.R. and south to Gabon, Congo-Kinshasa, Mozambique and Zimbabwe; widespread throughout the tropics

HAB. On lower parts of tree-trunks in moist forest, on rock in bamboo forest; 1250–2400 m

USES. None recorded

CONSERVATION NOTES. Widespread; least concern (LC)

SYN. *Trichomanes hirsutum* L., Sp. Pl.: 1098 (1753)

    Hymenophyllum boryanum Willd., Sp. Pl. 5: 518 (1810); Hook., Sp. Fil. 1: 89, pl. 31c (1844). Type: Réunion [Insula Borboniae], probably *Bory* s.n. (P!, holo.)

Sphaerocionium hirsutum (L.) C.Presl., Hymenophyllaceae: 34 (1834); Tardieu, Fl. Cam.
 Pterid.: 76 (1964); J.P.Roux, Conspect. southern Afr. Pteridoph.: 44 (2001)
Sphaerocionium boryanum (Willd.) C.Presl, Hymen.: 34, OR 126 (1843)
Hymenophyllum ciliatum Swartz in Schrad. Journ. 1800: 100 (1801); Syn. Fil.: 147 (1806);
 Christ. in Dansk Botanik. Arkiv. 7: 12 (1932); Copel., Hymen.: 169 (1937); Taton in Bull.
 Soc. Roy. Bot. Belg. 78: 17 (1946). Type: Jamaica, no type indicated
Sphaerocionium ciliatum (Swartz) C.Presl., Hymen.: 126 (1834)

4. **Hymenophyllum triangulare** *Bak.*, Syn. Fil.: 69 (1867); Tardieu in Fl. Camer.
Pter.: 74 (1964). Type: Bioko [Fernando Po], *Mann* 333 (K!, holo.)

Epiphyte; rhizome creeping, filiform, glabrous or with hairs to 1.5 mm. Fronds
spaced at 2–10 cm apart; stipe (2–)4–6 cm long, unwinged or with very narrow wings,
glabrous or with a few hairs; lamina triangular to oblong, 7–12 × 4–6 cm, rachis
narrowly winged, the wings with narrowly triangular lobes; pinnae 5–8, narrowly
ovate, to 3 × 2.5 cm; ultimate segments linear, to 4 × 1.5 mm, obtuse, margin
spinulose to denticulate; glabrous. Sori on the lower acroscopic lobe of the upper
pinnae, 2–3 mm long, 1.5–2 mm in diameter; indusial lobes ovate or elliptic,
glabrous; receptacle slightly exserted.

TANZANIA. Lushoto District: W Usambara Mts, Mazumbai, Apr. 1972, *Harris* 6236!; Morogoro
 District: Uluguru North, near Lupanga Peak, Dec. 1980, *Hall* s.n.!; Iringa District: Udzungwa
 Mountains National Park, peak 232, Oct. 2001, *Luke et al.* 8032!
DISTR. **T** 3, 6, 7; Cameroon, Bioko, Congo-Kinshasa, Rwanda, Burundi
HAB. Moist or mist forest, often on ridge-tops; low epiphyte; 1450–2100 m
USES. None recorded
CONSERVATION NOTES. Widespread; least concern (LC)

SYN. *Meringium triangulare* (Bak.) Copel. in Philip. Journ. Sc. 67: 44 (1938)
 Hymenophyllum triangulare Bak. subsp. *uluguruense* Kornaś in B.J.B.B. 60: 165 (1990), **syn.
 nov**. Type: Tanzania, N Uluguru Mts, Lupanga–Kinazi ridge, Nov. 1972, *Pócs &
 Mwanjabe* 6836/E (KRA, holo.; BR!, iso.)

NOTE. The subspecies was described by Kornaś as being 'slightly different' in several respects;
 the careful analysis of thirteen characters published by Kornaś and his co-authors shows
 overlap between all characters in the eastern and western populations in every case. The
 localities in Congo-Kinshasa, Rwanda and Burundi are geographically almost equidistant to
 the East African and the Cameroon/Bioko populations.

5. **Hymenophyllum peltatum** (*Poir.*) *Desv.* in Mem. Soc. Linn. Paris 6: 333 (1827);
Taton in Bull. Soc. Roy. Bot. Belg. 78: 20 (1946); Burrows, Southern African Ferns:
98, fig. 98, t. 20 (1990); J.P.Roux, Conspect. southern Afr. Pteridoph.: 41 (2001).
Type: Mauritius [Ile de France], *Bory de St Vincent* s.n. (P, holo., not found)

Epiphyte; rhizome creeping, thin, glabrous or nearly so. Fronds spaced 15 cm
apart, erect; stipe 7–30 mm long, glabrous, not winged or winged near apex;
lamina narrowly lanceolate-oblong, 2.5–10 × 1–2.1 cm, 2-pinnatifid; pinnae strongly
developed acroscopically; ultimate segments linear, to 1.3 mm wide, rounded,
margin serrate; glabrous on both surfaces. Sori usually 1 per pinna, on the innermost
acroscopic lobe, 1.8–2 mm long, 1.5–1.7 mm in diameter; valves entire, glabrous.
Fig. 2.11–2.13, p. 7.

UGANDA. Ruwenzori, Bujuku Valley near Bigo camp, Mar. 1948, *Hedberg* 432! & idem, near
 Nyamuleju Hut, Oct. 1971, *Pentecost* s.n.! & idem, Bujuku Valley 3 km above Mubuku bridge,
 Feb. 1997, *Lye & Katende* 22415!
TANZANIA. Kilimanjaro, Umbwe R. route, Mar. 1967, *Vesey-Fitzgerald* 5109!; S Pare Mts, Mt
 Shengena, Nov. 1986, *Schippers* 1626!; Morogoro District: Uluguru Mts, Bondwa summit, Apr.
 1974, *Faden & Faden* 74/408!
DISTR. **U** 2; **T** 2, 3, 6; Congo-Kinshasa (Ruwenzori), South Africa, Madagascar; widespread in
 temperate zones

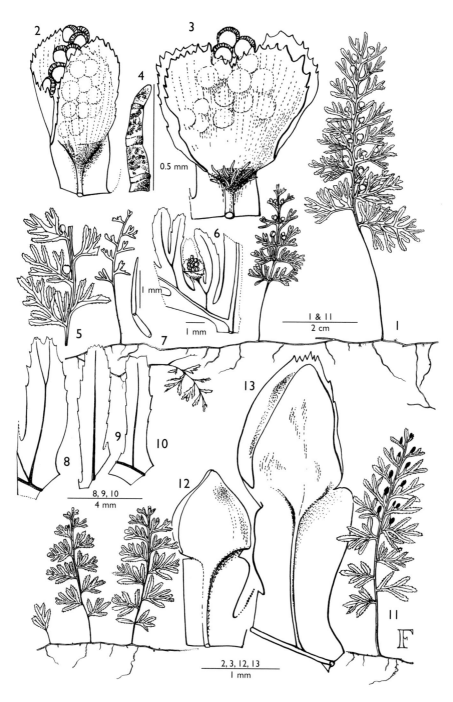

FIG. 2. *HYMENOPHYLLUM TUNBRIGENSE* — **1**, habit, × 1 ; **2, 3**, involucres of different ages, × 21; **4**, hair from base of involucre, × 55; **5**, pinnules, diagrammatic ; **6**, sorus, × 5; **7**, hair from stipe, × 50; **8–10**, ultimate segments, × 5. *HYMENOPHYLLUM PELTATUM* — **11**, habit, × 1; **12–13**, sori of different ages, × 21. 1–7, 10 from *Faden et al.* 69/516; 8 from *Faden & Holland* 71/182; 9 from *Drummond & Hemsley* 1770; 11–13 from *Faden* 74/408. Drawn by Monika Shaffer-Fehre.

HAB. Low epiphyte in moist forest or dense giant heath forest; 2100–3450 m
USES. None recorded
CONSERVATION NOTES. Widespread; least concern.

SYN. *Trichomanes peltatum* Poir. in Lam. Encycl. Meth. 8: 76 (1808)

6. **Hymenophyllum tunbrigense** (*L.*) *Sm.* in Sowerby, English Bot.: 3, t. 162 (1794);
Schelpe, F.Z. Pterid.: 80, t. 22E (1970); Burrows, S. Afr. Ferns: 97, fig., map (1990);
J.P.Roux, Conspect. southern Afr. Pteridoph.: 41 (2001). Type: 'habitat in Anglia,
Italia' (type not designated)

Epiphyte, less often on wet rocks; rhizome creeping, thin, glabrous or with a few
hair-like scales to 0.5 mm long. Fronds spaced 1–7 cm apart, erect, pale to dark
green; stipe 0.5–4 cm long, glabrous or with a few minute brown hairs, not winged or
winged in upper part; lamina ovate to lanceolate in outline, 3–14 × 1.2–3.5 cm, 2(–3)-
pinnatifid, rachis winged; pinnae in 6–14 pairs; ultimate segments 3–7 per pinna,
linear, to 0.9 mm wide, obtuse to truncate, margin serrate near apex; glabrous. Sori
2(–3) per pinna, on the inner acroscopic lobes, 1.1–2 mm long, 1.1–1.5 mm wide;
valves serrate. Fig. 2.1–2.10, p. 7.

KENYA. Meru District: Ithanguni, Kirui, Feb. 1970, *Faden & Evans* 70/84!; Embu District: Irangi
Forest Station, Apr. 1972, *Faden et al.* 72/182!; Kiambu District: Sasamua pipeline road,
Ndiara waterfall, Aug. 1974, *Faden et al.* 74/1338!
TANZANIA. Kilimanjaro: above Kidia, Dec. 1995, *Hemp* 934!; Lushoto District, Nov. 1974, *Balslev*
265!: Morogoro District: Bondwa Hill, Mar. 1953, *Drummond & Hemsley* 1770!
DISTR. K 4; T 2, 3, 6; Gabon, Malawi, Mozambique, Zimbabwe, South Africa; Madagascar and
Indian Ocean islands, western and southern Europe
HAB. Moist forest, ridge-top mist forests; low to medium epiphyte or on damp rocks in deep shade,
occasionally on overhanging earth banks; (1000–)1900–2700 m (to 2850 m fide Greenway)
USES. None recorded
CONSERVATION NOTES. Widespread; least concern.

SYN. *Trichomanes tunbrigense* L., Sp. Pl.: 1098 (1753)
Hymenophyllum thomassettii C.H.Wright in K.B. 1906: 170 (1906). Type: Malawi, Mt Mlanje,
Thomasset s.n. (K!, holo.)

USES. The specific epithet is sometimes spelled '*tunbridgense*' but the version without the 'd' is
the accepted name.

7. **Hymenophyllum capense** *Schrad.* in Gött. Gel. Anz. 1818: 919 (1818); Schelpe,
F.Z. Pterid.: 79 (1970); Burrows, S. Afr. Ferns: 95, fig. 20.95, t. 14.4, map (1990);
J.P.Roux, Conspect. southern Afr. Pteridoph.: 41 (2001). Type: South Africa, Cape of
Good Hope, *Hesse* s.n. (LE, holo.)

Low epiphyte; rhizome wide-creeping and branching, thin, glabrous or nearly so.
Fronds spaced 0.5–7 cm apart; stipe 0.4–7 cm long, narrowly winged in upper part,
glabrous; lamina narrowly ovate to lanceolate in outline, 1.5–11 × 1–3 cm, rather
irregularly 2–3-pinnatifid; pinnae somewhat acroscopically developed; pinnae
lobes up to 15 per pinna, to 1.5 mm wide, obtuse to shallowly emarginate, margin
entire; entire surface glabrous. Sori mainly on inside acroscopic lobes in upper
part of lamina, 1–2 per lobe apex; indusium ovate to obtriangular, soral valves
entire, 1.1–2 mm in diameter.

TANZANIA. Lushoto District: above Shume, Aug. 1986, *Schippers* 1497!; Morogoro District:
Nguru Mts near Mhonda Mission, Feb. 1971, *Mabberley & Pócs* 680!; Iringa District: Udzungwa
Mts above Sanje Falls, Nov. 1995, *de Boer et al.* 803!
DISTR. T 3, 6, 7; south to Malawi, Mozambique, Zimbabwe and South Africa; Madagascar
HAB. Low epiphyte in moist forest; 1450–2400 m
USES. None recorded
CONSERVATION NOTES. Widespread; least concern.

SYN. *Mecodium capense* (Schrad.) Pic.Serm. in Webbia 27: 404 (1972)

NOTE. Kornaś in B.J.B.B. 54: 15, t. 1, 2 (1984) lists his own collection, Morogoro District: Uluguru Mts, Bondwa Peak, Feb. 1972, *Kornaś* 1208C (KRA, not seen) as *Hymenophyllum inaequale* (Poir.) Desv. in Mém. Soc. Linn. Paris 6: 335 (1827). Type: Madagascar, collector unclear, 'herb. DC and du Petit Thouars'. Basionym: *Trichomanes inaequale* Poir. in Lam., Encycl. Méth., Bot. 8: 74 (1808); *Mecodium inaequale* (Poir.) Copel. in Philipp. J. Sci. 67: 96 (1938). The distribution of this taxon is Madagascar, Mascarene Islands, Comoros, Seychelles. Kornaś describes the differences with *H. kuhnii* but does not mention other taxa. As I have not seen the specimen I hesitate to make a decision about it but according to Kornaś' description the only differences with *H. capense* are the wingless stipe and wider lamina.

 Schlieben 3138! from Tanzania, Uluguru Mts NW, Dec. 1932 keys out here but is much shorter than *H. capense*. Fronds are at most 20 x 12 mm with simple, bifid or at most 4-lobed pinnae. Sori are 1-1.2 mm in diameter. I am uncertain about its status.

8. **Hymenophyllum kuhnii** C.Chr., Ind. Fil.: 363 (1905), *nom. nov.*; Taton in Bull. Soc. Roy. Bot. Belg. 78: 22 (1946); J.P.Roux, Conspect. southern Afr. Pteridoph.: 43 (2001). Types: Tanzania, Kilimanjaro, 1930–2800 m, *von Höhnel* 146 (B!, syn.) & 147 (B!, syn.) and *Ehlers* 66 (B!, syn.); 2500 m, Aug. 1888, *Meyer* s.n. (B!, syn.); Rua stream, *Meyer* 310 (B!, syn.). Lectotypus: *Meyer* 310 (B!, lecto. & isolecto., chosen here)

Epiphyte or lithophyte; rhizome creeping, filiform, glabrous or sometimes with a few brown hair-like scales to 1 mm. Fronds spaced 0.5–5 cm apart, the longer ones pendulous, dark to bright green; stipe 1–10 cm long, filiform, narrowly winged in upper part, glabrous or nearly so; lamina elongate-triangular to narrowly elliptic in outline, 5–50 × 2–6(–10) cm, 2–3-pinnatifid; pinnae to 8 × 1.5 cm; ultimate segments 3–5 on each side, linear, to 6 × 2.5 mm, rounded, margin entire; glabrous. Sori on the lower acroscopic lobe of the upper pinnae in the mid-part of the lamina, but also on upper ultimate segments in the more distal part of the lamina, 1.2–1.8 mm long and wide; indusial valves ovate, entire. Fig. 4: 1–4, p. 12.

UGANDA. Ruwenzori: Nyinabitaba, June 1953, *Osmaston* 3836! & idem, 1 km above Mubuku Bridge, Feb. 1997, *Lye* 22407!; Kigezi District: Bwindi forest, Ihihizo, Aug. 1998, *Hafashimana* 764!
KENYA. Kiambu District: Kieni forest, 8 km E of Kieni, June 1986, *Beentje & Mungai* 2904!; Kericho District: SW Mau forest along Kiptiget R., June 1972, *Faden et al.* 72/336!; Teita District: Kasigau, Bungule route, Nov. 1994, *Luke* 4186!
TANZANIA. Mt Meru, Jekukuma camp, Mar. 1968, *Greenway & Kanuri* 13280!; Lushoto District: Shume-Magamba Forest Reserve, May 1987, *Kisena* 492!; Morogoro District: Kanga Mt, Feb. 1970, *Pócs* 6139/U!
DISTR. **U** 2; **K** 3–5, 7; **T** 2, 3, 6, 7; from Sierra Leone to our area and south to South Africa
HAB. Moist forest, ridge-top mist forest, occasionally in giant heath forest but always in shady humid places; high to low epiphyte, on rock faces; may form large colonies; (900–)1400–3000 m
USES. None recorded
CONSERVATION NOTES. Widespread; least concern.

SYN. *Hymenophyllum meyeri* Kuhn in Engl., Hochgbfl. trop. Afr.: 94 (1892), *non* Presl (1843). Type as for *H. kuhnii*
 Trichomanes mildbraedii Brause in E.J. 53: 376 (1915). Type: Annobon, Santa Mina peak, *Mildbraed* 6701 (B!, holo. & iso.)
 Mecodium kuhnii (C.Chr.) Copel. in Philipp. Journ. Sci. 67: 19 (1938)
 Hymenophyllum mildbraedii (Brause & Hieron.) Alston in Catal. Vasc. Pl. São Tomé: 60 (1944)
 Hymenophyllum polyanthos (Sw.) Sw. var. *kuhnii* (C.Chr.) Schelpe in Bol. Soc. Brot. ser. 2: 40: 156 (1966) & in F.Z. Pterid.: 80 (1970); Burrows, S. Afr. Ferns: 97, fig., map (1990)
 Hymenophyllum polyanthos (Sw.) Sw. var. *mildbraedii* (Brause & Hieron.) Schelpe in Garcia de Orta Bot. 3, 1: 54 (1976)

NOTE. A 1935 collector whose name is ± illegible (?Turrabh?) describes this rather nicely as "delicate as green mist".
 An 1875 collection by *Cameron* gives as locality only 'Lake Tanganyika, Central Africa' which might have been **T** 1 or **T** 4, but it might have come from Zambia or Congo-Kinshasa as well (see Gillett in K.B. 14: 319).

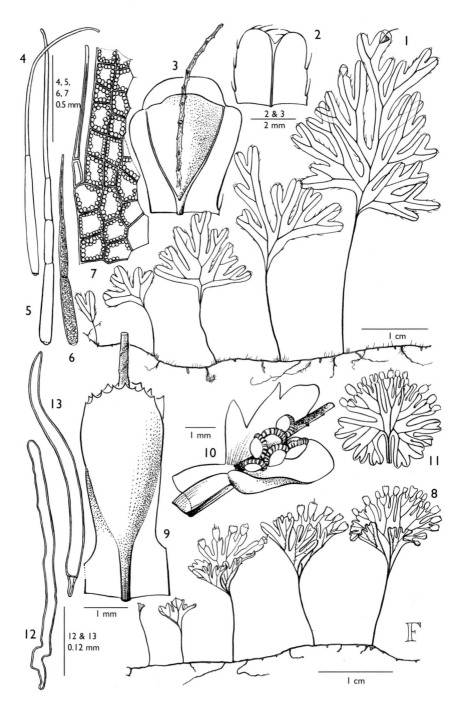

Fig. 3. *HYMENOPHYLLUM sp. A* — **1**, habit, × 1.7 ; **2**, pinnule apex, × 12; **3**, sorus, × 12; **4**, **5**, trichomes from rhizome, × 50; **6**, trichome from frond, × 50; **7**, lamina margin, × 50. *HYMENOPHYLLUM SIBTHORPIOIDES* — **8**, habit, × 2 ; **9**, sorus, × 12; **10**, opened indusium showing sporangia, × 8; **11**, frond, × 2; **12–13**, trichomes from frond abaxial surface, × 200. 1–7 from *Pócs & Nchimba* 6285/T; 8–13 from *Peter* 19369. Drawn by Monika Shaffer-Fehre.

9. **Hymenophyllum sibthorpioides** (*Willd.*) *Kuhn*, Filic. Afric.: 41 (1868); Schelpe, F.Z. Pterid.: 79 (1970); Burrows, S. Afr. Ferns: 94, fig. 95, t. 14/1, map (1990); J.P.Roux, Conspect. southern Afr. Pteridoph.: 43 (2001). Type: Reunion [Ins. Borboniae], *Bory de St Vincent* s.n. (P, holo., not found)

Low epiphyte; rhizome creeping, filiform, glabrous or with a few brown hairs to 0.6 mm. Fronds spaced 0.8–2.5 cm apart; stipe filiform, 5–32 mm long, glabrous or with a few brown hairs to 0.6 mm, unwinged; lamina dark green, fan-shaped to almost circular in outline, 6–22 mm in diameter, flabellately and palmately divided into up to 30 lobes; ultimate segments linear, to 4 × 1.5 mm, margin entire; veins repeatedly dichotomously branched; glabrous or with a few brown hairs to 0.6 mm on lower veins. Sori at lobe apex, 1–2 mm long, 1–1.5 mm in diameter; indusial valves ovate, erose-dentate. Fig. 3.8–3.13, p. 10.

KENYA. Teita District: Kasigau, June 1967, *Napper et al.* 2172! & idem, Rukanga route, Nov. 1994, *Luke & Luke* 4095D! & Mbololo forest, May 1985, *Beentje et al.* 974!
TANZANIA. Tanga District: E Usambara, Mlinga, July 1932, *Greenway* 3007!; Morogoro District: Nguru Mts near Maskati Mission, Mabega Mt, June 1978, *Thulin & Mhoro* 3045!; Iringa District: Udzungwa Mts, Mwanihana Forest reserve, Oct. 1984, *D.Thomas* 3865!
DISTR. **K** 7; **T** 3, 6, 7; Malawi, Mozambique, Zimbabwe; Madagascar, Comoro and Mascarene Islands
HAB. Moist forest, mist forest; low epiphyte or on shady rocks; 850–1950 m
USES. None recorded
CONSERVATION NOTES. Widespread; least concern.

SYN. *Trichomanes sibthorpioides* Willd., Sp. Pl. ed. 4, 5: 498 (1810)

10. **Hymenophyllum sp. A** (*Pócs* 6285/T)

Lithophyte; rhizome creeping, filiform, with a few brown simple trichomes to 0.8 mm long. Fronds spaced 0.3–3 cm apart; stipe filiform, 6–24 mm long, glabrous or with a few simple trichomes, unwinged; lamina fan-shaped or broadly ovate in outline, 15–35 × 15–24 mm, irregularly divided to almost 3-pinnatifid with up to 27 lobes; ultimate segments linear and up to 9 × 2.3 mm, margin entire with a few dark marginal trichomes to 0.3 mm long, apex rounded or emarginate; veins repeatedly dichotomously branched; glabrous but for the marginal trichomes. Sori at lobe apex, 1.5–2.3 mm long, to 1.8 mm in diameter, indusial valves obovate, entire; receptacle exserted up to 2 mm. Fig. 3: 1–7, p. 10.

TANZANIA. Morogoro District: Uluguru Mts, Bondwa, Nov. 1932, *Schlieben* 3028b! & idem, N Uluguru Mts, SW ridge of Lupanga above Mbete, Nov. 1970, *Pócs & Nchimbi* 6285/T! & idem, Jan. 1986, *Pócs* 8612/D! & Nguru Mts above Mhonda, Jan. 1987, *Schippers* 1707!
DISTR. **T** 6
HAB. On moist shady rocks in moist forest; 1700–1900 m
USES. None recorded
CONSERVATION NOTES. Data deficient (DD) but possibly at least vulnerable – forest at this altitude is disappearing due to agricultural expansion

NOTE. Another specimen, Uluguru Mts, Bondwa, Apr. 1974, *Faden & Faden* 74/389! looks very similar but lacks the marginal trichomes – see Fig. 4.5–4.7, p. 12.
The Berlin website has *Schlieben* 3028 from the Uluguru Mts as the type of *Trichomanes digitatum* Sw. var. *uluguruense* Reimers, as far as I can tell an unpublished name. The specimen looks quite like *Pócs* 6285/T, but is mixed with *Hymenophyllum splendidum*.

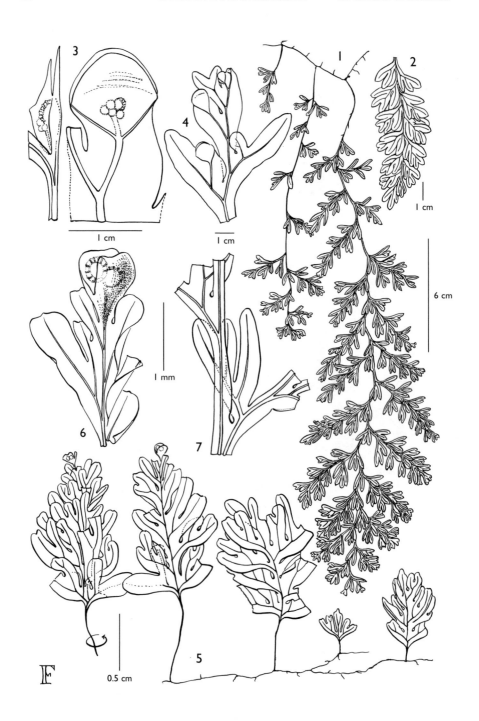

Fig. 4. *HYMENOPHYLLUM KUHNII* — **1**, habit, × 0.7 ; **2**, pinna, × 1.1; **3**, involucre, laterally and from top, × 20; **4**, pinna apex, × 5. *HYMENOPHYLLUM sp.* (see note sub sp. A) — **5**, habit, × 3 ; **6**, pinna apex, × 12; **7**, lamina part; × 18. 1–4 from *Winkler* 3912; 5–8 from *Faden & Faden* 74/389. Drawn by Monika Shaffer-Fehre.

2. DIDYMOGLOSSUM

Desv., Prodr.: 330 (1827); Ebihara et al. in Blumea 51, 2: 235–237 (2006)

Trichomanes L. sect. *Didymoglossum* (Desv.) T.Moore, Ind. Fil.: 110 (1857)
Trichomanes L. subgen. *Didymoglossum* (Desv.) C.Chr., Ind. Fil.: 14 (1906)

Rhizome long-creeping, usually filiform, densely hairy; roots absent, root-like shoots present. Fronds spaced; lamina simple to pinnatifid; venation catadromous, often flabelliform; submarginal vein present (subgen. *Microgonium*) or absent (subgen. *Didymoglossum*), longitudinal false veins present. Sori often immersed in lamina, campanulate, lips bilabiate or less often truncate; receptacle exserted.

More than 30 species; throughout the tropics.

1. Marginal vein present in lamina; lamina margin glabrous 2
   Marginal vein absent; sori unwinged; lamina margin with
   simple stiff hairs to 0.6 mm . . . .   1. *D. reptans*
2. Lamina entire to irregularly lobed; sori > 1.5 mm long,
   narrowly winged, the mouth dilated . . . .   2. *D. erosum*
   Lamina pinnately lobed to bipinnate; sori 1–1.5 mm long,
   unwinged, mouth not dilated . . . .   3. *D. lenormandii*

1. **Didymoglossum reptans** (*Sw.*) *C.Presl*, Hymenophyllaceae: 23 (1843). Type: Jamaica, *Swartz* s.n. (S, holo.; BM, SBT, iso.)

Low epiphyte or lithophyte; rhizome creeping, with dark brown linear-lanceolate hairs to 1 mm long. Fronds spaced 0.3–2 cm apart; stipe 1–7 mm long, with hairs as on the rhizome; lamina oblong to narrowly obovate, 1–5.5 × 0.4–2.6 cm, very irregularly pinnatifid, margin sparsely set with pairs of stiff dark brown hairs to 0.6 mm; marginal vein absent. Sori at lobe apices of distal lobes; indusium narrowly obconical, 2–3.3 mm long, 0.6–1 mm in diameter, not or hardly winged, lips flared and to 6 mm long and 13 mm wide. Fig. 5, p. 14.

TANZANIA. Morogoro District: Uluguru Mts, Morningside–Bondwa, July 1970, *Faden et al. 70/335!* & Nguru Mts, Turiani, Mhonda, Jan. 1987, *Schippers* 1677!; Iringa District: Udzungwa Mts above Sanje waterfall, Nov. 1995, *de Boer et al.* 761!
DISTR. **T** 6, 7; South Africa, Madagascar; Central and South America
HAB. Moist forest along stream; low epiphyte; 1250–1350 m
USES. None recorded
CONSERVATION NOTES. Least concern; widespread

SYN. *Trichomanes reptans* Sw., Nova Gen. Sp. Pl.: 136 (1788); Burrows, S. Afr. Ferns: 89, fig., map (1990); J.P.Roux, Conspect. southern Afr. Pteridoph.: 46 (2001)

NOTE. Resembles *D. erosum* but differs in ± free sorus, plus marginal lamina hairs and the lack of marginal vein.

2. **Didymoglossum erosum** (*Willd.*) *Beentje* **comb. nov.** Type: 'Oware et Benin', *Flugge* s.n. (B-W 20189, holo.)

Lithophyte, low epiphyte or terrestrial; rhizome wide-creeping, filiform, ± branched, with dense red-brown to dark brown crisped hairs to 1 mm long. Fronds spaced at 0.3–5 cm intervals; stipe 0.1–2.5 cm long, with hairs like those of rhizome when young; lamina dark green, rarely pale or bright green, simple, elliptic, oblong-ovate or obovate, 1–6 × 0.3–3 cm, entire or irregularly crenate to irregularly lobed, base attenuate; indument of simple or stellate hairs when young, especially on midrib and margins; midrib strong, pinnately or flabellately branched in larger fronds,

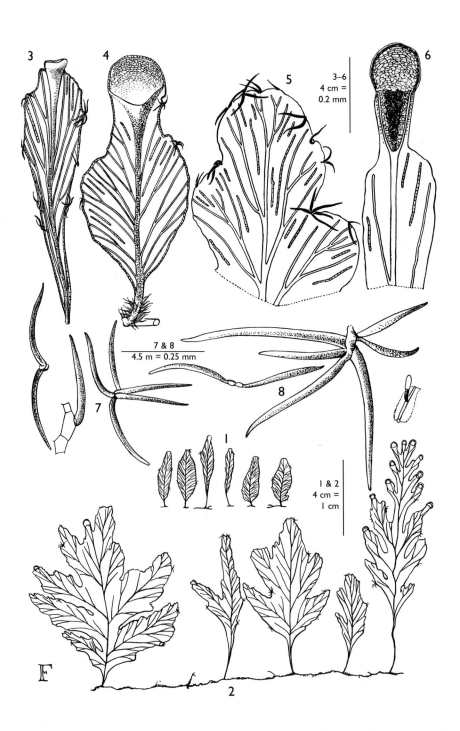

FIG. 5. *DIDYMOGLOSSUM REPTANS* — **1**, lamina shapes, × ¹/₂; **2**, habit, × 2; **3, 4**, lamina with sorus, × 10; **5**, lamina margin, × 10 ; **6**, lamina apex, × 10; **7, 8**, marginal trichomes, × 90. 1–4 from *Swartz* s.n. (type); 3–4 from *Holttum* 28315. Drawn by Monika Shaffer-Fehre.

submarginal vein present; false veinlets many. Sori marginal, in upper part of leaf, up to 9; indusium slender-fusiform to urceolate, 1.5–2.5(–4) mm long, 0.3–0.7 mm in diameter, narrowly winged, mouth dilated to 1.5 mm diameter but hardly bilabiate; receptacle exserted slightly or up to 4 mm when old. Fig. 6, p. 16.

UGANDA. Kigezi District: Ishasha R. 7 km SW of Kirima, Sep. 1969, *Lye et al.* 4190!; Ankole District: Kalinzu forest 4 km NW of saw mill, Sep. 1969, *Faden et al.* 69/1151!; Masaka District: Sesse Islands, 1904, *Dawe* 50!

KENYA. Meru District: Ithanguni Forest 17 km from Nkubu, base of Kirui, June 1969, *Faden et al.* 69/769!; Kiambu District: Kikuyu escarpment forest, Ndiara waterfall along Sasamua pipeline road, Aug. 1974, *Faden et al.* 74/1336!; Kwale District: Shimba Hills, Kivumoni, Mar. 1991, *Luke & Robertson* 2696!

TANZANIA. Lushoto District: East Usambara Mts, Hunga Valley NW of Amani, Nov. 1986, *Borhidi & Pócs* 86/594!; Morogoro District: Nguru South Forest reserve above Kwamanga, near Mhonda Mission, Feb. 1971, *Mabberley & Pócs* 703!; Iringa District: Udzungwa Mountain National Park, Sonjo–Mwanihana route, Nov. 1997, *Luke & Luke* 4993b!; Pemba: Ngezi Forest, Dec. 1930, *Greenway* 2706!

DISTR. U 2, 4; K 4, 7; T 2, 3, 6, 7; P; from Sierra Leone in the west to C.A.R. and Kenya, and south to South Africa; Madagascar, Comoro Is., Seychelles

HAB. On shaded rock, low on tree-trunks (often on *Cyathea*) or on moist stream-banks; always in moist forest, usually near water; may be locally common and form mats; 0–2400 m

USES. None recorded

CONSERVATION NOTES. Least concern (LC); widespread

SYN. *Trichomanes erosum* Willd., Sp. Pl. 4th ed., 5: 501 (1810); Taton in Bull. Soc. Roy. Bot. Belg. 78: 28 (1946); Tardieu in Mém. I.F.A.N. 28: 41, t. 4.3–4 (1955); Alston, Ferns W.T.A.: 30 (1959); Tardieu, Fl. Camer. 3: 82 (1964); Schelpe in F.Z. Pteridoph.: 76 (1970); Burrows, S. Afr. Ferns: 89, fig., map (1990); J.P.Roux, Conspect. southern Afr. Pteridoph.: 48 (2001)
   Microgonium erosum (Willd.) C.Presl. in Abh. Boehm. Ges. Wiss. 5, 5: 335 (1848)
   Trichomanes aerugineum Bosch in Nederl. Kruidk. Arch. ser. 1, 5(3): 201 (1863); Taton in Bull. Soc. Roy. Bot. Belg. 78: 26 (1946); Tardieu in Mém. I.F.A.N. 28: 43 (1955); Alston, Ferns W.T.A.: 30 (1959). Type: Bioko [Fernando Po], *Barter* s.n. (K!, holo.)
   T. erosum Willd. var. *aerugineum* (Bosch) R.Bonap. in Notes Pterid. 13: 165 (1929)
   T. erosum Willd. var. *majus* Peter, F.D.-O.A.: 12 (1929) & (as forma *majus*) Descr.: 1 (1929). Type: Tanzania, Usambara Mts, Dodwe valley near Amani, *Peter* 14467 (?B, holo., not found)
   Trichomanes chamaedrys Taton in Bull. Soc. Roy. Bot. Belg. 78: 29 (1946); Alston, Ferns W.T.A.: 30 (1959). Type: Congo-Kinshasa, Mbandaka [Coquilhatville], *Dewèvre* 580 (BR!, holo.)
   T. erosum Willd. var. *chamaedrys* (Taton) Tardieu in Mém. I.F.A.N. 28: 42, t. 4.5 (1955)
   Crepidomanes erosum (Willd.) Kunkel in Nova Hedwigia 6: 211 (1963)
   C. chamaedrys (Taton) Kunkel in Nova Hedwigia 6: 212 (1963)
   Microgonium aerugineum (Bosch.) Pic.Serm. in Webbia 23: 181 (1968)
   M. chamaedrys (Taton) Pic.Serm. in Webbia 23: 181 (1968)

NOTE. Several treatments maintain two varieties as follows:

Lamina entire to crenate; sori mainly in upper half . . . .   var. *erosum*
Lamina pinnatifid; sori all over . . . .   var. *aerugineum*
   but intermediates occur and the 'varieties' can sometimes be found together on a single specimen. I treat them as synonymous.

3. **Didymoglossum lenormandii** (*Bosch*) *Ebihara & Dubuisson* in Blumea 51, 2: 236 (2006). Type: Comoro Is., Mayotte, *Boivin* s.n. (P, holo., not found)

Low epiphyte or terrestrial; rhizome creeping, filiform, with dense brown hairs to 0.5 mm. Fronds spaced 0.2–1.2 cm apart; stipe 1–20 mm long, winged in the upper half, glabrous except for base; lamina lanceolate or deltoid, (1–)2–4.5 × 0.5–3 cm, pinnately lobed to bipinnate; pinnae 2–5 mm wide, obtuse, glabrous; pinnately veined, intramarginal vein present but not very distinct, often interrupted; false veins many. Sori many, lateral, one per lobe; indusium narrowly turbinate, 1–1.5 mm long, 0.4–0.5 mm in diameter, mouth not dilated; receptacle exserted for up to 12 mm, often recurved. Fig. 7, p. 17.

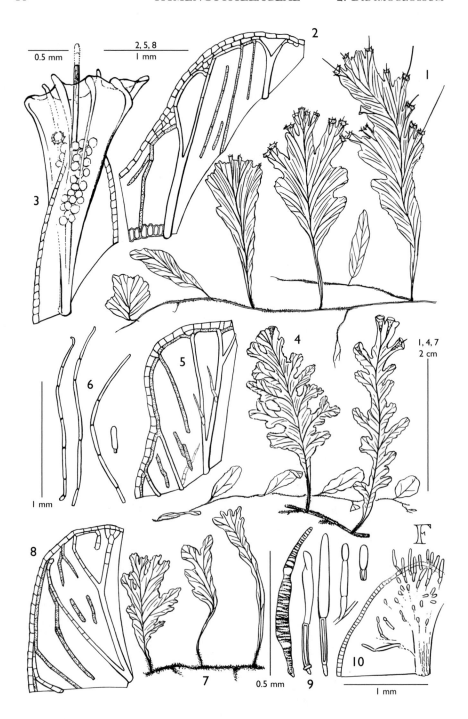

Fig. 6. *DIDYMOGLOSSUM EROSUM* — **1**, habit, × 1.7 ; **2**, section of frond showing true and false (dotted) veins, × 40; **3**, sorus, × 22; **4**, habit, × 1.7; **5**, frond margin, × 40 ; **6**, scales from frond apex, × 65; **7**, habit, × 1.7; **8**, frond margin, × 40; **9**, hairs and scales from frond apex, × 65; **10**, young frond apex, × 22. 1–3 from *Mabberley* 1360; 4–7 from *Lye* 4190; 8–10 from *Faden* 71/140. Drawn by Monika Shaffer-Fehre.

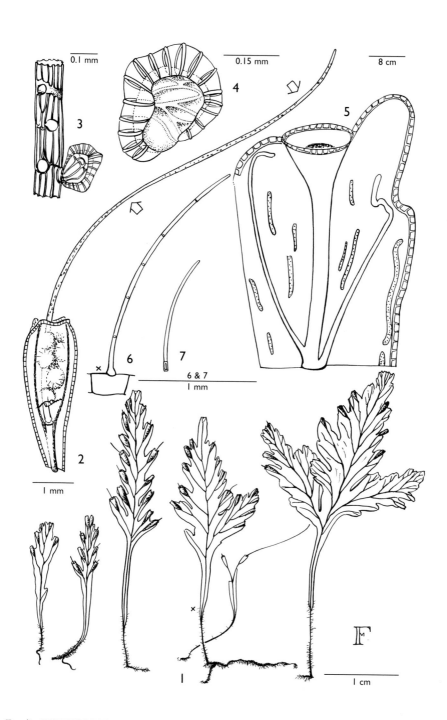

FIG. 7. *DIDYMOGLOSSUM LENORMANDII* — **1**, frond shapes from a single specimen, × 2; **2**, sorus, × 11; **3**, detail of sporangiophore, much enlarged; **4**, sporangium, × 11; **5**, frond margin with sorus; **6**, **7**, hairs from stipe/lamina decurrence. All from *Kornaś* 978. Drawn by Monika Shaffer-Fehre.

TANZANIA. Lushoto District: Usambara Mts, Dodwe stream near Amani, June 1970, *Faden 70/269*! & Kwamkoro Forest reserve bordering Kwamsambia Forest Reserve, May 1987, *Borhidi et al. 87/238*! ; Morogoro District: Nguru South Forest Reserve, near Mhonda Mission, Feb. 1971, *Mabberley & Pócs 671*!

DISTR. **T** 3, 6; Madagascar

HAB. Moist forest by stream, where regularly collected as epiphytic on *Cyathea* or *Marattia*, and once collected from roadside bank; 650–1200 m

USES. None recorded

CONSERVATION NOTES. Data deficient. Not many collections from our area, but distribution in Madagascar unclear.

SYN. *Trichomanes lenormandii* Bosch in Nederl. Kruidk. Arch. ser. 1, 5(2): 144 (1861); Kornaś in B.J.B.B. 46: 393, t. 1, 2 (1976)
   Microgonium lenormandii (Bosch) Copel. in Philipp. Journ. Sci. 67: 104 (1938)

3. **CREPIDOMANES**

(C.Presl.) C.Presl, Epimeliae bot.: 258 (1851); Ebihara et al. in Blumea 51, 2: 237–240 (2006)

Trichomanes L. subgen. *Crepidomanes* C.Presl, Epimeliae bot.: 17 (1851)

Plants epilithic or epiphytic. Rhizome widely creeping, irregularly branched, with a few scattered roots and adhesive hairs; indument of simple uniseriate acicular hairs. Fronds widely spaced; stipe terete, often distally winged; lamina 1-layered, flabellate, digitate or pinnately compound; ultimate segments glabrous or setiferous when young; rachis winged; venation free, anadromous, with or without false veinlets. Indumentum of unicellular naviculate secretory hairs along the veins, simple hairs on frond axes. Sori terminal on the veins, solitary at the apex of the ultimate segments, involucre cup-shaped with bilabiate mouth; receptacle long and extruding.

120 species, mostly in the tropics of the Old World.

1. Lamina irregularly divided or palmately flabellate 2
   Lamina 2–5-pinnatifid 4
2. Stipe sometimes proliferous (visible as small 'bud' or
      as small plant growing out of stipe, usually in quite
      a few fronds per plant), not winged, with quite a
      few trichomes; lamina cells about twice as long as
      broad; indusium 1.6–1.8 mm long, with flared
      mouth to 1.5 mm diameter .... 3. *C. mannii*
   Stipe never proliferous 3
3. Stipe winged, usually quite glabrous; lamina 10–25
      × 10–25 mm, lamina cells about as long as broad;
      indusium 1–1.5 mm long, with flared mouth to
      1.5 mm diameter .... 1. *C. chevalieri*
   Stipe winged, with a few scattered trichomes; lamina
      usually much longer than wide; indusium 2–2.5 mm
      long, mouth not dilated .... 2. *C. ramitrichum*
4. Stipe sometimes proliferous (visible as small 'bud' or
      as small plant growing out of stipe, usually in quite
      a few fronds per plant), not winged, with quite a
      few trichomes; lamina cells about twice as long as
      broad; indusium 1.6–1.8 mm long, with flared
      mouth to 1.5 mm diameter .... 3. *C. mannii*
   Stipe never proliferous 5

5. Some of the rhizome hairs branched; indusium 2-
lobed, not dilated at the mouth ....   2. *C. ramitrichum*
All rhizome hairs simple; indusium dilated at the
mouth, not 2-lobed 6
6. Sori sunk into lobe, to 0.7 mm in diameter ....   4. *C. fallax*
Sori mostly free, wider at mouth 7
7. Longitudinal drying folds present; sori 1.5–2.5 mm
long, to 2 × as long as wide, narrowly winged (lobe
narrowing at base of involucre) ....   5. *C. melanotrichum*
Longitudinal drying folds absent; sori 2–2.5 mm
long, 2–3 × as long as wide, broadly winged (lobe
not narrowed anywhere) ....   *Polyphlebium borbonicum*

1. **Crepidomanes chevalieri** (*Christ*) *Ebihara & Dubuisson* in Blumea 51, 2: 238 (2006). Type: Central African Republic, Krebedjé, Tomi valley, *Chevalier* 5400 (P, holo.; K!, fragm.)

Low epiphyte or occasionally a lithophyte; rhizome filiform, creeping, with dense unbranched brown hairs to 0.5 mm long. Fronds spaced at 0.5–2 cm intervals; stipe 2–25 mm long, glabrous except for the base, near apex with some minute scales when young; lamina dark green, flabelliform, irregularly and rather deeply divided into 5–6 main divisions, 1–2.5 × 1–2.5 cm; ultimate segments linear, 1–2 mm long and up to 0.8 mm wide, obtuse to slightly emarginate; false veins absent; when dry with longitudinal pleats; lamina cells 1–1.8(–2.4) × as long as wide; with some minute scales when young. Sori terminal on segments in lower and middle part of lamina, less often in upper part as well; indusium obconic to urceolate, 1–1.5 mm long, 0.5–1 mm in diameter, slightly winged, the lips flared to 1.5 mm diameter; receptacle exserted to 2.5 mm. Fig. 9: 1–4, p. 21 .

UGANDA. Bunyoro District: Budongo Forest, second stream crossing Sanso saw mill road, Sep. 1969, *Faden et al.* 69/1087!; Toro District: Kibale Forest National Park, 13 km E of Fort Portal, Sep. 1997, *Lye & Katende* 22929!; Mengo District: Kajansi forest, km 16 Entebbe road, July 1937, *Chandler* 1784!

KENYA. North Kavirondo District: Kakamega Forest, forest station, Apr. 1965, *Gillett* 16689!; Machakos/Masai District: Chyulu Hills, main forest camp 3, Oct. 1997, *Luke & Luke* 4814!; Kwale District: Shimba Hills National Park, Sheldrick's Falls, Nov. 1970, *Faden et al.* 70/812!

TANZANIA. Arusha District: Ngongongare, Sep. 1951, *Greenway & Hughes* 8561!; Lushoto District: Derema near Amani, Hunga R., Mar. 1974, *Faden et al.* 74/364!; Kigoma District: Ulembe–Ikoba track 97 km from Ikola, Nov. 1959, *Richards* 11727!

DISTR. **U** 2–4; **K** 4–7; **T** 2–4, 6; Guinea to C.A.R. and Kenya and south to Congo-Kinshasa and Tanzania

HAB. Moist forest, a low epiphyte or less often on spray-zone rocks; (150–)500–2400 m

USES. None recorded

CONSERVATION NOTES. Least concern (LC); widely distributed

SYN. *Trichomanes chevalieri* Christ in Bull. Soc. Bot. France 55, Mém. 8: 106 (1908); Alston, Ferns W.T.A.: 30 (1959); Tardieu, Fl. Camer. 3: 85, t. 10.3–4 (1964); Benl, Pter. Fernando Po: 20 (1980); Kornaś in Fragm. Florist. Geobot 39 : 51, fig. 13 (1994)
Vandenboschia chevalieri (Christ) Kunkel in Nova Hedwigia 6: 213 (1963)

2. **Crepidomanes ramitrichum** (*R.B.Faden*) *Beentje* **comb. nov.** Type: Kenya, Kericho District: SW Mau Forest, Kipteget R., June 1972, *Faden & Grumbley* 72/338 (EA!, holo., B, BM, BOL, BR!, DSM, GH, K!, LISC, LMU, MHU, MO, P, PRÉ, SRGH, US, WAG, iso.)

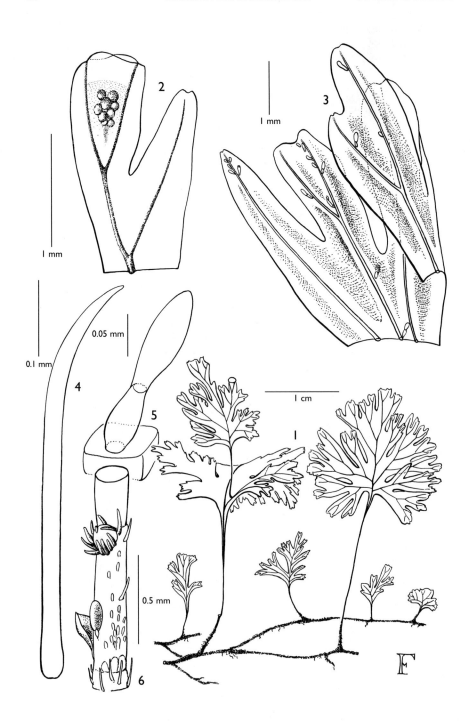

FIG. 8. *CREPIDOMANES MANNII* — **1**, habit, × 2 ; **2**, frond lobe apex with sorus, × 30; **3**, lamina margin, × 14; **4**, trichome from stipe base, × 100; **5**, glandular hair, × 240 ; **6**, part of stipe, × 47. From *Paget-Wilkes* 889 and *Kornaś* 1756. Drawn by Monika Shaffer-Fehre.

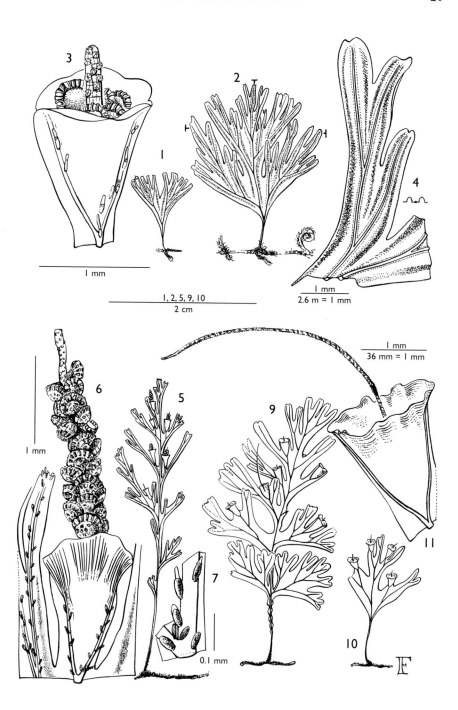

FIG. 9. *CREPIDOMANES CHEVALIERI* — **1**, **2**, habit; **3**, sorus; **4**, lamina part showing folds. *CREPIDOMANES MELANOTRICHUM* — **5**, habit; **6**, sorus; **7**, trichomes on vein. *CREPIDOMANES METTENII* — **9**, **10**, habit; **11**, sorus. 1, 3–4 from *Faden & Faden* 73/164; 2 from *Lucas* 111; 5–7 from *Tweedie* 2969; 9–11 from *Andrews* 249 from Sudan. Drawn by Monika Shaffer-Fehre.

Low epiphyte or lithophyte; rhizome filiform, covered in dark brown trichomes to 1.5 mm, some of these branched. Fronds spaced up to 2 cm apart; stipe dark green, blackish at base, 2–25 mm long, winged in upper part, with scattered minute clavate hairs; lamina ovate to linear in outline, 2–14 × 0.8–3 cm, flabellately divided or 2–3-pinnatifid, ultimate segments linear; longitudinal drying folds present; glabrous except for some minute hairs on the veins. Sori terminal on basal acroscopic pinnule lobes, sometimes also on (sub)terminal pinna lobes; indusium tubular, 2–2.5 mm long, winged, mouth bilabiate, usually not spreading; receptacle exserted. Fig. 10: 3–5, p. 23.

UGANDA. Toro District: Itwara forest near Sogoli R., Jan. 1997, *Lye & Katende* 22170!; Kigezi District: Ishasha Gorge, Sep. 1969, *Faden* 69/1194! & Impenetrable Forest, Sep. 1961, *Rose* 10262!

KENYA. North Kavirondo District: Malava forest, W side of Kakamega–Broderick Falls road, Nov. 1969, *Faden & Evans* 69/2045a!; Kericho District: Marinyn Tea Estate, 4 km SSE of Kericho, June 1972, *Faden et al.* 72/325!; Teita District: Msau–Mbololo road, Mwambirwa Forest Station, Sep. 1970, *Faden et al.* 70/524!

TANZANIA. Lushoto District: Derema, Hunga R., Mar. 1974, *Faden & Faden* 74/362! & Gonja, Bulua, Sep. 1893, *Holst* 4288!; Morogoro District: Mindu Mts, Apr. 1970, *Pócs* 6152/U!

DISTR. **U** 2; **K** 5, 7; **T** 3, 6; Congo-Kinshasa, Rwanda, Burundi, Zambia, Mozambique, Zimbabwe

HAB. Moist forest especially near water, or rock faces by falls; may be locally common; 750–2150 m

USES. None recorded

CONSERVATION NOTES. Widespread; least concern (LC)

SYN. *T. borbonicum sensu* Schelpe, F.Z. Pterid.: 76 (1970), pro parte
T. pyxidiferum L. var. *melanotrichum sensu* Schelpe, F.Z. Pterid.: 78 (1970), pro parte, *non* (Schltdl.) Schelpe
T. sp. A of U.K.W.F. ed. 1: 29 (1974)
Trichomanes ramitrichum R.B.Faden in Amer. Fern J. 67(1): 5, fig. 1–5 (1977)

NOTE. Fronds closely resemble those of *C. melanotrichum* and *Polyphlebium borbonicum*. This taxon differs in the branched trichomes on the rhizome and the shape of the indusium. I am not sure whether this actually warrants specific status – but I am also unwilling to meddle with the status quo based on slight doubts!

A Mascarene taxon is rather close: **Crepidomanes frappieri** (*Cordem.*) *J.P.Roux*, Conspect. southern Afr. Pteridoph.: 45 (2001). Type: Reunion, sine loco, sine collectore (not found). Neotype: Reunion, Brule de St Denis, 1891, *Bedier* s.n. (P!, holo.), chosen by Pic.Serm.; synonym *Vandenboschia frappieri* (Cordem.) Pic.Serm. in Webbia 37: 131 (1983). As its basionym is *Trichomanes frappieri* Cordem. in Bull. Soc. Sci. Arts Reunion 1890–91: 143 (1891); Kornaś in Fragm. Florist. Geobot 39 : 62, fig. 23 (1994), this epithet would have priority. Again, I am unwilling to meddle with the status quo.

3. **Crepidomanes mannii** (*Hook.*) *J.P.Roux*, Conspect. southern Afr. Pteridoph.: 45 (2001). Type: Bioko [Fernando Po], *Mann* s.n. (K!, holo.)

Low epiphyte or lithophyte, rarely terrestrial; rhizome long-creeping, filiform, branched, with dense light to dark brown hairs to 0.8 mm. Fronds spaced 0.5–5 cm apart; stipe 10–25 mm long, sometimes proliferous, not winged, glabrous except at the very base; lamina dark green, suborbicular to ovoid or rhomboid, irregular, 1–5(–9) × 1–3 cm, palmately or flabellately and dichotomously dissected, the segments sometimes themselves pinnatifid, occasionally 2-pinnatifid, base cuneate; ultimate segments ± 0.5 mm wide, obtuse or emarginate; with longitudinal folds on drying; lamina cells 2–4 × as long as wide; minute scales present when young. Sori up to 8 in number, marginal, often axillary in the sinuses between the lobes in the upper part of the lamina; indusium turbinate to campanulate, 1.5–1.8 mm long, 0.6–0.8 mm in diameter, minutely winged, mouth widened to 1 mm, wavy, scarcely 2-lipped; receptacle often exserted. Fig. 8, p. 20.

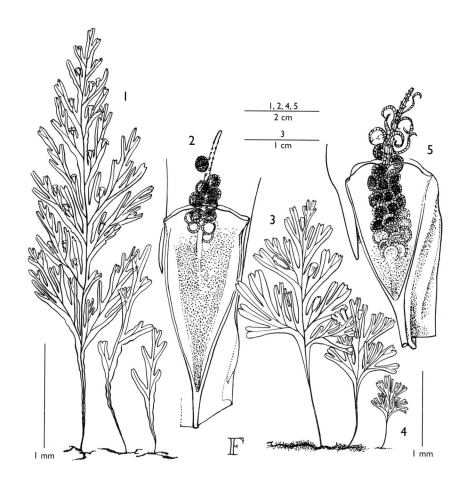

Fig. 10. *CREPIDOMANES FALLAX* — **1**, habit, × 1 ; **2**, sorus, × 22. *CREPIDOMANES RAMITRICHUM* — **3**, habit, × 2 ; **4**, habit, × 1; **5**, pinnule apex with sorus, × 20. 1–2 from *Wood* Y49; 3–5 from *Faden et al.* 70/524. Drawn by Monika Shaffer-Fehre.

UGANDA. Kigezi District: Ishasha Gorge, 6 km SW of Kirima, Sep. 1969, *Faden et al.* 69/1195!; Mt Elgon, Oct. 1960, *Rose* 10268!; Mengo District: 1.5 km NE of Nansagazi, Sep. 1969, *Faden et al.* 69/1026!

KENYA. Kericho District: Tea Research Institute 5 km E of Kericho, June 1972, *Faden et al.* 72/313!

TANZANIA. Lushoto District: W Usambara, Shagayu Forest Reserve, Kwashemhambu, Oct. 1986, *Borhidi et al.* 86/054!; Morogoro District: Kanga Mts, Nov. 1970, *Pócs* 6140/N!; Iringa District: Udzungwa Mountain National Park, Mwaya–Mwanihana route, Nov. 1997, *Luke & Luke* 4912!

DISTR. **U** 2–4; **K** 5; **T** 2, 3, 6, 7; Sierra Leone to Sudan and south to Angola and Zambia; Madagascar, Comoro and Mascarene Islands

HAB. Moist forest, where often by streams; on wet rocks, tree trunks, rarely on soil banks; may be mat-forming; 750–2150 m

USES. None recorded

CONSERVATION NOTES. Least concern (LC); widely distributed

SYN. *Trichomanes mannii* Hook., Syn. Fil.: 75 (1867); Taton in Bull. Soc. Roy. Bot. Belg. 78: 30
     (1946); Alston, Ferns W.T.A.: 30 (1959); Kornaś in B.J.B.B. 46: 387, t. 1–3 (1976) & in
     Fragm. Florist. Geobot 39 : 43, fig. 7a–b (1994)
   Trichomanes ruwenzoriensis Taton in Bull. Soc. Roy. Bot. Belg. 78: 31 (1946); Kornaś in
     Fragm. Florist. Geobot 39 : 45, fig. 9a–b (1994), as *ruwenzoriense*. Type: Congo-Kinshasa,
     Ruwenzori, Lamia, 2000 m, *Bequaert* s.n. (BR!, holo.), **syn. nov.**
   Gonocormus mannii (Hook.) Kunkel in Nova Hedwigia 6: 212 (1963)

4. **Crepidomanes fallax** (*Christ.*) *Ebihara & Dubuisson* in Blumea 51, 2: 238 (2006).
Type: Congo-Kinshasa, Zobia R., Buta, *Seret* 863 (BR!, holo.).

Epiphyte or lithophyte; rhizome creeping, much branched, with dense dark brown
hairs to 0.3 mm long. Fronds spaced up to 35 mm apart, glabrous; stipe 3–30 mm long,
winged in upper part, glabrous except for very base; lamina grey-green to dark green,
narrowly lanceolate, 6–16 × 2–3.5 cm, bipinnatifid; base attenuate; rachis broadly
winged; pinnae 6–13 on each side, lanceolate, to 2.2 × 0.5 cm, decurrent on rachis,
ascending, divided in up to 7 segments; ultimate segments to 12 × 2 mm, obtuse to
slightly emarginate. Sori marginal, axillary or terminal, in mid- and upper part of
lamina, always sunk into the lobe; indusium turbinate to campanulate, 1.5–2 mm long,
to 0.8 mm in diameter, the mouth dilated and to 1.5 mm in diameter; receptacle
exserted to 5 mm. Fig. 10: 1–2, p. 23.

UGANDA. Masaka District: Bugala Island, Towa Forest, July 1939, *A.S. Thomas* 3020! & June 1950,
     Wood Y49!
DISTR. U 4; Guinea to Ghana, Fernando Po, Congo-Kinshasa; Madagascar
HAB. Moist forest, as an epiphyte on tree trunks and *Uapaca* roots; ± 1140 m
USES. None recorded
CONSERVATION NOTES. Least concern (LC); widespread, though very rare in Uganda.

SYN. *Trichomanes fallax* Christ. in Ann. Mus. Congo ser. 5, 3: 24 (1909); Taton in Bull. Soc. Roy.
     Bot. Belg. 78: 36 (1946); Tardieu in Mém. I.F.A.N. 28: 43, t. 4.7–8 (1955); Alston, Ferns
     W.T.A.: 31 (1959); Benl, Pter. Fernando Po: 21 (1980); Kornaś in Fragm. Florist. Geobot
     39 : 57, fig. 19 (1994)
   Vandenboschia fallax (Christ) Copel. in Philipp. J. Sci. 67: 52 (1938)

5. **Crepidomanes melanotrichum** (*Schltdl.*) *J.P.Roux*, Conspect. southern Afr.
Pteridoph.: 46, t. 5h–i (2001). Type: South Africa, Cape Peninsula, Plettenberg Bay,
± 1825, *Mundt & Maire* s.n. (B!, holo.; HAL 37269, P, iso., not found)

Lithophyte or epiphyte; rhizome creeping, branched, filiform, with dense black or
very dark brown hairs to 1 mm long. Fronds spaced 0.3–4.5 cm apart, sometimes
pendulous but more often erect; stipe 5–35 mm long, compressed, narrowly winged
in upper half, glabrous except for base or with almost pectinate hairs on lower wing
margins; lamina usually dark green, narrowly oblong to narrowly ovate, 0.8–14 ×
0.8–3.5 cm, 2–3-pinnatifid; rachis winged; pinnae to 15 on each side, to 2.2 × 1 cm,
folded longitudinally when drying; pinnules 5–7, ultimate lobes to 8 × 1 mm,
rounded; veins with minute brown scales. Sori 1–7 per pinna, throughout the
lamina, mostly axillary, sometimes bent above the plane of the leaf; indusium
conical to campanulate, 1.5–2(–2.5) mm long, 1–1.2 mm in diameter, narrowly
winged or not winged, mouth dilated, not 2-lipped; receptacle exserted to 2(–6) mm.
Fig. 9: 5–7, p. 21.

UGANDA. Karamoja District: Kokumongole, Awitok R., May 1939, *A.S. Thomas* 2902!; Bunyoro District: 9.5 km N of Kabwoya, Sep. 1969, *Faden et al.* 69/1093!; Mengo District: Sezibwa Falls, Aug. 1938, *Chandler* 2457!

KENYA. Trans Nzoia District: Cherangani Hills, Kaibwibich, Aug. 1968, *Thulin & Tidigs* 55!; Kiambu District: 10 km E of Kieni, June 1986, *Beentje* 2950!; Teita District: Maungu Hills, S of Maungu Station, May 1970, *Faden et al.* 70/172!

TANZANIA. Mt Meru, E footslopes, Mar. 1968, *Greenway & Kanuri* 13370!; Kigoma District: Kungwe Mt, July 1959, *Newbould & Harley* 4767!; Morogoro District: Uluguru Mts, above Morningside, July 1970, *Faden et al.* 70/336!

DISTR. **U** 1–4; **K** 1–7; **T** 2–4, 6–8; Sierra Leone to Ethiopia and south to South Africa; Madagascar and Mascarene Islands

HAB. Moist forest and then in moist and shady sites, occasionally in riverine or drier type forest; on tree-trunks, moist rock-faces, mossy rocks, rarely on moist earth banks; may be locally common; (?200–)750–2650 m

USES. None recorded

CONSERVATION NOTES. Least concern; widespread

SYN. *Trichomanes melanotrichum* Schltdl., Adumbr.: 56 (1825) & in Linnaea 10: 553 (1836); Tardieu in Mém. I.F.A.N. 28: 46, t. 5.1–2 (1955) & Fl. Camer. 3: 87, t. 10.7–8 (1964); Burrows, S. Afr. Ferns: 92, fig., map (1990); Kornaś in Fragm. Florist. Geobot 39 : 64, fig. 25 (1994)

Trichomanes pyxidiferum sensu Taton in Bull. Soc. Roy. Bot. Belg. 78: 32 (1946), *non* L.

Trichomanes pyxidiferum L. forma *major* Taton in Bull. Soc. Roy. Bot. Belg. 78: 34 (1946). Type: Congo-Kinshasa, Ruwenzori, *Bequaert* 3931b (BR, syn.), 4231 (BR, syn.), s.n. (twice, BR, syn.), 4245 (BR!, syn.), **syn. nov.**

Vandenboschia melanotricha (Schltdl.) Pic.Serm. in Webbia 12: 127 (1956)

Trichomanes pyxidiferum L. var. *melanotrichum* (Schltdl.) Schelpe in Journ. S.Afr. Bot. 30: 181 (1964) & F.Z. Pterid.: 78 (1970); Benl, Pter. Fernando Po: 24 (1980)

Trichomanes africanum sensu Alston, Ferns W.T.A.: 31 (1959), *non* Christ

Vandenboschia inopinata Pic.Serm. in B.J.B.B. 53: 245 (1983). Type: Congo-Kinshasa, Mt Shamulamba, *Pichi Sermolli* 4455 (Herb. Pic.Serm., holo.; BR!, iso.)

Trichomanes inopinatum (Pic.Serm.) Burrows, S. Afr. Ferns: 93, fig., map (1990); Kornaś in Fragm. Florist. Geobot 39 : 68, fig. 27 (1994)

Trichomanes inopinatum (Pic.Serm.) Burrows var. *majus* (Taton) Kornaś in Fragm. Florist. Geobot 39 : 71, fig. 29 (1994). Type as for *Trichomanes pyxidiferum* L. forma *major* Taton

NOTE. A specimen from Kenya, Shimba Hills (*van Someren* 114, Mar. 1941) has no altitude indicated, but must be from a lower altitude than all other material collected.

Burrows and Kornaś recognize *Trichomanes inopinatum* (Pic.Serm.) Burrows, close to *melanotrichum* but with differences listed below:

melanotrichum rhizome hairs black, shiny; lower part of stipe wing with black, articulated hairs; stipe winged all along

inopinatum rhizome hairs dark brown to blackish; stipe glabrous to sparsely set with brown clavate hairs [true]; stipe winged in upper part only [true]

I have seen the type, and I prefer to treat this as a synonym rather than as a taxon in its own right. There are minute differences but 'form' is the most I would agree to.

Similarly *Trichomanes pyxidiferum* forma *major* of Taton = *Trichomanes inopinatum* var. *majus* of Kornaś, is hardly different.

4. POLYPHLEBIUM

Copel. in Philip. Journ. Sc. 67: 55 (1938); Ebihara et al. in Blumea 51,
2: 240–241 (2006)

Phlebiophyllum Bosch in Versl. Meded. Afd. Natuurk. Kon. Akad. Wetensch. 11: 321
(1861), *non Phlebophyllum* Nees (1832)
Trichomanes L. subgen. *Polyphlebium* (Copel.) Allan, Fl. New Zealand 1: 34 (1961)

Rhizomes long-creeping, filiform, densely hairy; roots few and fine. Fronds spaced;
lamina pinnate to quadripinnate, venation anadromous, sometimes with a single row of
marginal cells, false veinlets absent. Sori tubular, lips usually dilate; receptacle exserted.

15 species in southern temperate regions and montane forests of the tropics

Polyphlebium borbonicum (*Bosch*) *Ebihara & Dubuisson* in Blumea 51, 2: 238
(2006). Type: Réunion [Ins. Borboniae], *Boivin* 908 (B!, lecto.; P, iso.)

Lithophyte or occasionally a low epiphyte; rhizome creeping, filiform, with
unbranched brown to black hairs to 1 mm long. Fronds spaced at 1–4 cm intervals;
stipe 6–65 mm long, narrowly winged in upper half, glabrous except near very base;
lamina ovate to lanceolate in outline, 2.5–14 × 1.5–5.5 cm, 2–3-pinnatifid; rachis
winged; pinnae 6–14 on each side, up to 3.5 × 1 cm, spreading at 45–90°; ultimate
segments linear, obtuse to emarginate; false veins or longitudinal drying folds
absent. Sori in upper part of lamina, indusium narrowly obconic, 2–2.5 mm long,
0.6–1.4 mm in diameter, (narrowly) winged; lips flared, receptacle exserted up to
7 mm. Fig. 11: 1–4, p. 27.

UGANDA. Kigezi District: Maramagambo [Malambigambo] forest, Sep. 1961, *Rose* 10330!
KENYA. Nyandarua/Aberdare Mts, Sasamua Dam, Jan. 1970, *Faden et al.* 71/81!; Mt Kenya, first
    bridge above Thiba Fishing Camp, Oct. 1979, *Gilbert* s.n. (range 5785–5796)!; Teita District:
    Kasigau, Bungule route, Nov. 1994, *Luke* 4187!
TANZANIA. Kilimanjaro, above Kidia, Dec. 1995, *Hemp* 899!; Morogoro District: Uluguru Mts,
    Mwere Valley, Sep. 1970, *Faden et al.* 70/590! & Mwanihana Forest Reserve above Sanje, Oct.
    1984, *D.W. Thomas* 3868!
DISTR. **K** 3, 4, 7; **T** 2, 6; from Guinea to Kenya and S to South Africa; Madagascar, Mascarene
    Islands
HAB. In moist forest by streams, either on moss-covered rocks or (less often) a low epiphyte,
    once described as terrestrial; usually uncommon, but sometimes locally common and on
    Ithanguni said to be locally dominant; 1400–2600 m
USES. None recorded
CONSERVATION NOTES. Least concern (LC); widely distributed

SYN. *Trichomanes borbonicum* Bosch in Nederl. Kruidk. Arch. ser. 1, 5(2): 158 (1861); Alston,
    Ferns W.T.A.: 31 (1959); Schelpe in F.Z. Pteridoph.: 76 (1970); Benl, Pter. Fernando Po:
    22 (1980); Burrows, S. Afr. Ferns: 93, t. 13.7, map (1990); Kornaś in Fragm. Florist.
    Geobot 39 : 60, fig. 21 (1994)
    Trichomanes goetzei Hieron. in E.J. 28: 339 (1900); Taton in Bull. Soc. Roy. Bot. Belg. 78: 34,
    t. 3a–b (1946). Type: Tanzania, S Uluguru Mts, *Goetze* 194 (B!, holo.)
    Vandenboschia borbonica (Bosch) Kunkel in Nova Hedwigia 6: 213 (1963)
    Crepidomanes borbonicum (Bosch) J.P.Roux, Conspect. southern Afr. Pteridoph.: 45 (2001)

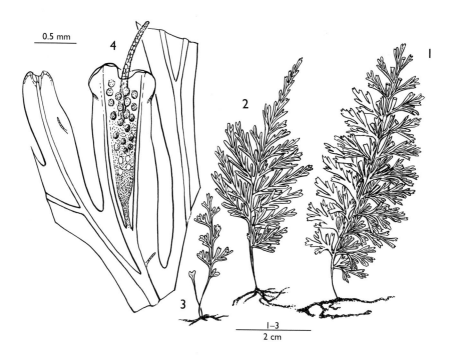

FIG. 11. *POLYPHLEBIUM BORBONICUM* — **1–3**, habit, × 1 ; **4**, pinnule apex with sorus, × 22. 1–4 from *Faden et al.* 70/607 & 70/590. Drawn by Monika Shaffer-Fehre.

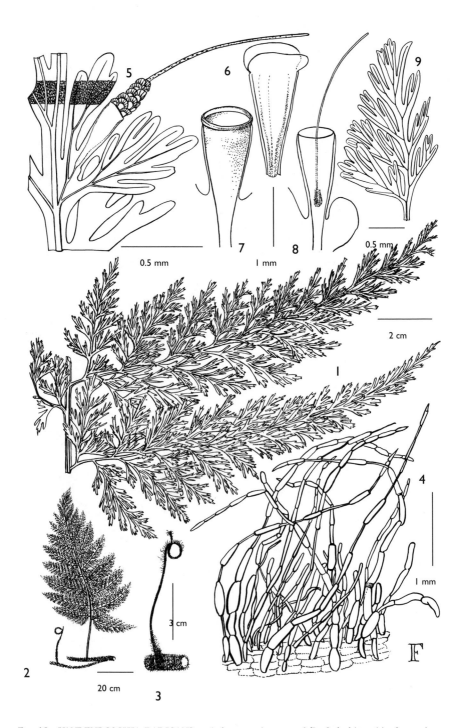

FIG. 12. *VANDENBOSCHIA RADICANS* — **1**, largest pinnae, × 0.7 ; **2**, habit, × ¹/₂₀; **3**, crozier, × 0.5; **4**, hairs, × 20; **5**, pinnule apex with sorus, × 4.5; **6, 7, 8**, involucre, × 16; **9**, pinnule, × 4. 1–2 from *Faden et al.* 70/609; 3–4, 9 from *Thomas* 3682; 5–7 from *Lovett & Congdon* 602; 8 from *Faden et al.* 69/1221. Drawn by Monika Shaffer-Fehre.

5. **VANDENBOSCHIA**

Copel. in Philip. Journ. Sc. 67: 51 (1938); Ebihara et al. in Blumea 51,
2: 241–242 (2006)

Trichomanes L. subgen. *Vandenboschia* (Copel.) Allan, Fl. New Zealand 1: 34 (1961)

Rhizome short- or long-creeping, rather thick, densely hairy; roots numerous and
robust. Fronds spaced; lamina 2–5-pinnatifid, venation anadromous, false veinlets
absent. Sori tubular to campanulate, lips sometimes dilated; receptacle exserted.

15 species throughout the tropics, also in northern temperate regions.

Vandenboschia radicans (*Sw.*) *Copel.* in Philipp. J. Sci. 67: 54 (1938).   Type:
Jamaica, mountains, *Swartz* s.n. (S, holo.)

Low epiphyte, epilithic or occasionally terrestrial; rhizome creeping and up to 2 m
long, 1–5 mm in diameter, with abundant rootlets and with dense dark articulate
hairs to 3 mm long. Fronds spaced 2–10 cm apart; stipe erect, 3–17 cm long, terete,
winged throughout or only near the lamina, glabrous or with sparse hairs; lamina
dark green, broadly ovate to ovate, 15–50 × 9–20 cm, 4(–5)-pinnatifid; rachis winged
or not winged, with sparse hairs; pinnae alternate, up to 22 on each side of the rachis,
petiolulate, falcate, to 14 × 4 cm; pinnules up to 15 pairs, to 3.5 × 2 cm; costae and
costules winged, with occasional minute brown scales; ultimate segments linear, to
1 mm wide, obtuse. Sori 1–4 per pinnule, sub-axillary; involucre urceolate to
narrowly campanulate, stalked to immersed, 1.5–2 mm long, to 0.5 mm in
diameter, mouth truncate or dilated, very slightly bilabiate; receptacle exserted
for up to 6 mm. Fig. 12, p. 28.

var. **radicans**

UGANDA. Ankole District: Kasyoha-Kitomi Forest reserve, NE of Kyambura R., June 1994,
Poulsen et al. 562!; Kigezi District: Ishasha gorge, Nov. 1946, *Purseglove* 2254! & Nov. 1997,
Hafashimana 472!
KENYA. Fort Hall District: Thika Falls, Nov. 1968, *Faden* 68/822a!
TANZANIA. Kilimanjaro, Mrusunga valley between Natiro and Uru, Dec. 1999, *Hemp* 2443!;
Lushoto District: W Usambara Mts, Baga II Forest reserve above Kambi Falls, Jan. 1985, *Pócs
& Temu* 85/433!; Iringa District: Udzungwa escarpment, Lulando Forest, Apr. 1986, *Lovett &
Congdon* 602!
DISTR. **U** 2; **K** 4; **T** 2, 3, 6, 7; Nigeria, Gulf of Guinea islands, east to our area and south to
Angola and Congo-Kinshasa; pantropical
HAB. Moist forest, usually near stream, once in riverine forest by waterfall; on tree trunks, wet
rock, or occasionally terrestrial; 1000–1800 m
USES. None recorded for our area
CONSERVATION NOTES. Least concern (LC); widespread

SYN. *Trichomanes radicans* Sw. in J. Bot. (Schrader) 1800(2): 97 (1801); Benl, Pter. Fernando Po:
14 (1980); Kornaś in Fragm. Florist. Geobot 39 : 54, fig. 15 (1994)
Trichomanes giganteum Willd., Sp. Pl. 4th ed., 5: 514 (1810); Tardieu in Mém. I.F.A.N. 28: 45
(1955); Alston, Ferns W.T.A.: 31 (1959). Type: Réunion, [Ins. Borboniae], *Bory de St
Vincent* s.n. (not found)
Crepidomanes radicans (Sw.) K.Iwats. in J. Fac. Sci. Univ. Tokyo, Bot. 13: 530 (1985);
J.P.Roux, Conspect. southern Afr. Pteridoph.: 45 (2001)

NOTE. The other variety, var. **naseanum** (Christ) K.Iwats. occurs in Asia.

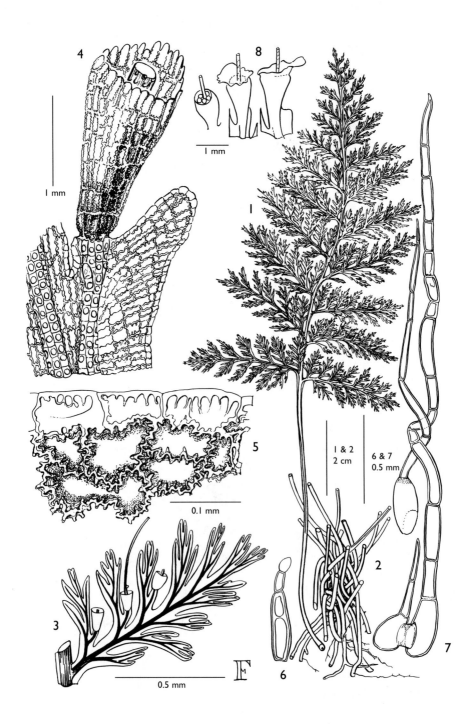

FIG. 13. *ABRODICTYUM RIGIDUM* — **1**, frond, × 1; **2**, rhizome, × 1; **3**, pinnule, × 5; **4**, involucre, × 50; **5**, pinnule margin, × 200 ; **6**, **7**, glandular hair and trichomes, × 100; **8**, involucres of varying age, × 30. 1–8 from *Greenway* 4170; 9 from *Faden et al.* 70/984. Drawn by Monika Shaffer-Fehre.

6. ABRODICTYUM

Presl, Hymenophyllum: 20 (1843); Ebihara et al. in Blumea 51, 2: 242–244 (2006)

Cephalomanes C.Presl. subgen. *Abrodictyum* (C.Presl.) K.Iwats. in Acta Phytotax. Geobot. 35(4-6): 176 (1984)

Rhizomes erect or short-creeping, thick, densely hairy; roots numerous, robust. Fronds clustered or somewhat spaced; lamina 2-pinnate to 4-pinnatifid, venation anadromous, false veinlets absent. Sori tubular, lips truncate; receptacle exserted.

25 species; pantropical.

Abrodictyum rigidum (*Sw.*) *Ebihara & Dubuisson* in Blumea 51, 2: 243 (2006). Type: Jamaica, sine coll. s.n. (S, holo., LD, UPS, iso.)

Terrestrial or lithophyte; rhizome short-creeping or erect, to 3 mm in diameter, with dense brown trichomes to 1 mm long. Fronds tufted, erect, dark green; stipe brown, 1.5–22 cm long, narrowly winged near apex, glabrous or with sparse brown multicellular hairs to 1.5 mm long; lamina ovate to lanceolate in outline, 4–24 × 2–14 cm, 3–4-pinnatifid; rachis narrowly winged in upper part of lamina; pinnae 14–20 on each side, shortly stalked, to 7 × 3 cm; pinnules to 13 on each side, to 1.7 × 0.5 cm; ultimate segments linear, to 0.3 mm wide, attenuate, glabrous on both surfaces. Sori proximal on the acroscopic side of pinnules, rarely a few on the basiscopic pinnules as well, usually 1 per pinnule, close to costules; indusium narrowly conical, 0.7–1.5 mm long, 0.5–0.7 mm in diameter, not winged, mouth dilated to 0.9 mm, valves entire; receptacle exserted for up to 7 mm. Fig. 13, p. 30.

UGANDA. Kigezi District: Ishasha R. 7 km SW of Kirima, Sep. 1969, *Faden et al.* 69/1226! & Sep. 1969, *Lye et al.* 4188!; Masaka District: Sesse Is., Bugala Island, Jungo forest near Mweno, Sep. 1997, *Lye & Katende* 22774!
KENYA. Teita District: Mt Kasigau, Apr. 1969, *Faden et al.* 69/467! & idem, above Bungule, Dec. 1970, *Faden et al.* 70/984!
TANZANIA. Pare District: Chome Forest reserve, Shengena Mts, Dec. 1998, *Hemp* 2259!; Tanga District: Kwamkoro, Dec. 1959, *Semsei* 2958!; Morogoro District: N Uluguru Forest Reserve, Mnavu, Dec. 1993, *Kisena* 990!
DISTR. **U** 2, 4; **K** 7; **T** 3, 6, 7; West Africa from Guinea to Kenya and south to South Africa; Madagascar, Seychelles, tropical America
HAB. Moist forest where it occurs in dark and wet places such as deeply shaded stream banks, caves, under large rocks; terrestrial or less often lithophytic; may be locally common; 850–2000 m
USES. None recorded for our area
CONSERVATION NOTES. Least concern; widespread

SYN. *Trichomanes rigidum* Sw., Prodr. (Swartz): 137 (1788), *non* Hedw. 1802 *nec* Wall. 1828 *nec* Hk. & Baker 1867; Taton in Bull. Soc. Roy. Bot. Belg. 78: 37 (1946); Schelpe, F.Z. Pterid.: 78, t. 22A (1970); Burrows, S. Afr. Ferns: 90, fig., map (1990); Kornaś in Fragm. Florist. Geobot 39 : 38, fig. 3 (1994)
Selenodesmium rigidum (Sw.) Copel. in Philip. Journ. Sc. 67: 81 (1938)
Trichomanes cupressoides Desv., Prod. Foug.: 330 (1827); Alston, Ferns W.T.A.: 31 (1959); Tardieu in Fl. Camer. Pter.: 90 (1964). Type: Seychelles, herb. Desvaux sine coll. s.n. (P, holo., not found)
Cephalomanes rigidum (Sw.) K.Iwats. in Acta Phytotax. Geobot. 35: 177 (1984); J.P.Roux, Conspect. southern Afr. Pteridoph.: 48 (2001)
NOTE. Many older specimens were named *Trichomanes mandioccanum* Raddi, which is an Asian species.
Taton in Bull. Soc. Roy. Bot. Belg. 78: 37 (1946) states *T. rigidum* sensu Peter, F.D.-O.A.: 14 (1929) is not this taxon, but he does not say which taxon it represents.

INDEX TO HYMENOPHYLLACEAE

32

New names validated in this part

Crepidomanes ramitrichum (*R.B.Faden*) *Beentje* **comb. nov.**
Didymoglossum erosum (*Willd.*) *Beentje* **comb. nov.**

GEOGRAPHICAL DIVISIONS OF THE FLORA

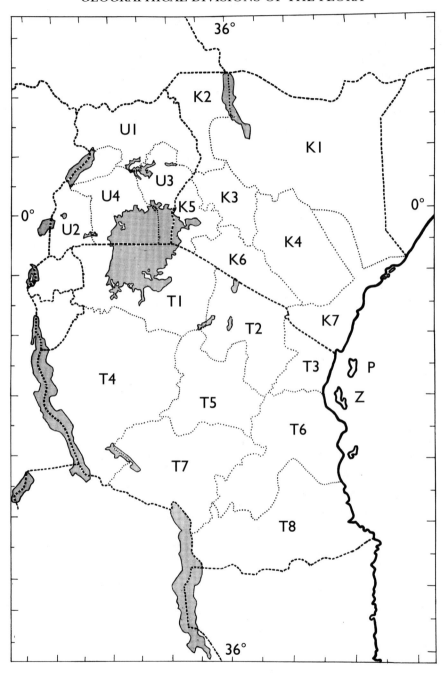

First published in 2008 by
Royal Botanic Gardens, Kew
Richmond, Surrey, TW9 3AB, UK
www.kew.org

ISBN 978 1 84246 373 4

British Library Cataloguing in Publication Data
A catalogue record for this book is available from the British Library

Design and typesetting by Margaret Newman,
Kew Publishing, Royal Botanic Gardens, Kew.

For information or to purchase all Kew titles please visit
www.kewbooks.com or email publishing@kew.org

All proceeds go to support Kew's work in saving the world's plants for life

LIST OF ABBREVIATIONS

A.V.P. = O. Hedberg, Afroalpine Vascular Plants; **B.J.B.B.** = Bulletin du Jardin Botanique de l'Etat, Bruxelles; Bulletin du Jardin Botanique Nationale de Belgique; **B.S.B.B.** = Bulletin de la Société Royale de Botanique de Belgique; **C.F.A.** = Conspectus Florae Angolensis; **E.J.** = A. Engler, Botanische Jahrbücher für Systematik, Pflanzengeschichte und Pflanzengeographie; **E.M.** = A. Engler, Monographieen Afrikanischer Pflanzen-Familien und Gattungen; **E.P.** = A. Engler, Das Pflanzenreich; **E.P.A.** = G. Cufodontis, Enumeratio Plantarum Aethiopiae Spermatophyta; in B.J.B.B. 23, Suppl. (1953) et seq.; **E. & P. Pf.** = A. Engler & K. Prantl, Die Natürlichen Pflanzenfamilien; **F.A.C.** = Flore d'Afrique Centrale (*formerly* F.C.B.); **F.C.B.** = Flore du Congo Belge et du Ruanda-Urundi; Flore du Congo, du Rwanda et du Burundi; **F.E.E.** = Flora of Ethiopia & Eritrea; **F.D.-O.A.** = A. Peter, Flora von Deutsch-Ostafrika; **F.F.N.R.** = F. White, Forest Flora of Northern Rhodesia; **F.P.N.A.** = W. Robyns, Flore des Spermatophytes du Parc National Albert; **F.P.S.** = F.W. Andrews, Flowering Plants of the Anglo-Egyptian Sudan *or* Flowering Plants of the Sudan; **F.P.U.** = E. Lind & A. Tallantire, Some Common Flowering Plants of Uganda; **F.R.** = F. Fedde, Repertorium Speciorum Novarum Regni Vegetabilis; **F.S.A.** = Flora of Southern Africa; **F.T.A.** = Flora of Tropical Africa; **F.W.T.A.** = Flora of West Tropical Africa; **F.Z.** = Flora Zambesiaca; **G.F.P.** = J. Hutchinson, The Genera of Flowering Plants; **G.P.** = G. Bentham & J.D. Hooker, Genera Plantarum; **G.T.** = D.M. Napper, Grasses of Tanganyika; **I.G.U.** = K.W. Harker & D.M. Napper, An Illustrated Guide to the Grasses of Uganda; **I.T.U.** = W.J. Eggeling, Indigenous Trees of the Uganda Protectorate; **J.B.** = Journal of Botany; **J.L.S.** = Journal of the Linnean Society of London, Botany; **K.B.** = Kew Bulletin, *or* Bulletin of Miscellaneous Information, Kew; **K.T.S.** = I. Dale & P.J. Greenway, Kenya Trees and Shrubs; **K.T.S.L.** = H.J. Beentje, Kenya Trees, Shrubs and Lianas; **L.T.A.** = E.G. Baker, Leguminosae of Tropical Africa; **N.B.G.B.** = Notizblatt des Botanischen Gartens und Museums zu Berlin-Dahlem; **P.O.A.** = A. Engler, Die Pflanzenwelt Ost-Afrikas und der Nachbargebiete; **R.K.G.** = A.V. Bogdan, A Revised List of Kenya Grasses; **T.S.K.** = E. Battiscombe, Trees and Shrubs of Kenya Colony; **T.T.C.L.** = J.P.M. Brenan, Check-lists of the Forest Trees and Shrubs of the British Empire no. 5, part II, Tanganyika Territory; **U.K.W.F.** = A.D.Q. Agnew (or for ed. 2, A.D.Q. Agnew & S. Agnew), Upland Kenya Wild Flowers; **U.O.P.Z.** = R.O. Williams, Useful and Ornamental Plants in Zanzibar and Pemba; **V.E.** = A. Engler & O. Drude, Die Vegetation der Erde, IX, Pflanzenwelt Afrikas; **W.F.K.** = A.J. Jex-Blake, Some Wild Flowers of Kenya; **Z.A.E.** = Wissenschaftliche Ergebnisse der Deutschen Zentral-Afrika-Expedition 1907–1908, 2 (Botanik).

FAMILIES OF VASCULAR PLANTS REPRESENTED IN
THE FLORA OF TROPICAL EAST AFRICA

The family system used in the Flora has diverged in some respects from that now in use at Kew and the herbaria in East Africa. The accepted family name of a synonym or alternative is indicated by the word "see". Included family names are referred to the one used in the Flora by "in" if in accordance with the current system, and "as" if not. Where two families are included in one fascicle the subsidiary family is referred to the main family by "with".

PUBLISHED PARTS

Foreword and preface
*Glossary
Index of Collecting Localities

Acanthaceae
  Part 1
*Actiniopteridaceae
*Adiantaceae
Aizoaceae
Alangiaceae
Alismataceae
*Alliaceae
*Aloaceae
*Amaranthaceae
*Amaryllidaceae
*Anacardiaceae
*Ancistrocladaceae
Anisophyllaceae — as Rhizophoraceae
Annonaceae
*Anthericaceae
Apiaceae — see Umbelliferae
Apocynaceae
  *Part 1
*Aponogetonaceae
Aquifoliaceae
*Araceae
Araliaceae
Arecaceae — see Palmae
*Aristolochiaceae
Asparagaceae
*Asphodelaceae
Aspleniaceae
Asteraceae — see Compositae
Avicenniaceae — as Verbenaceae
*Azollaceae

*Balanitaceae
*Balanophoraceae

*Balsaminaceae
Basellaceae
Begoniaceae
Berberidaceae
Bignoniaceae
Bischofiaceae — in Euphorbiaceae
Bixaceae
Blechnaceae
*Bombacaceae
*Boraginaceae
Brassicaceae — see Cruciferae
Brexiaceae
Buddlejaceae — as Loganiaceae
*Burmanniaceae
*Burseraceae
Butomaceae
Buxaceae

Cabombaceae
Cactaceae
Caesalpiniaceae — in Leguminosae
*Callitrichaceae
Campanulaceae
Canellaceae
Cannabaceae
Cannaceae — with Musaceae
Capparaceae
Caprifoliaceae
Caricaceae
Caryophyllaceae
*Casuarinaceae
Cecropiaceae — with Moraceae
*Celastraceae
*Ceratophyllaceae
Chenopodiaceae
Chrysobalanaceae — as Rosaceae
Clusiaceae — see Guttiferae
Cobaeaceae — with Bignoniaceae
Cochlospermaceae

Papaveraceae
Papilionaceae — in Leguminosae
*Parkeriaceae
Passifloraceae
Pedaliaceae
Periplocaceae — see Apocynaceae (Part 2)
Phytolaccaceae
*Piperaceae
Pittosporaceae
Plantaginaceae
Plumbaginaceae
Poaceae — see Gramineae
Podocarpaceae
Podostemaceae
Polemoniaceae — see Cobaeaceae
Polygalaceae
Polygonaceae
*Polypodiaceae
Pontederiaceae
*Portulacaceae
Potamogetonaceae
Primulaceae
*Proteaceae
*Psilotaceae
*Ptaeroxylaceae
*Pteridaceae

*Rafflesiaceae
Ranunculaceae
Resedaceae
Restionaceae
Rhamnaceae
Rhizophoraceae
Rosaceae
Rubiaceae
  Part 1
  *Part 2
  *Part 3
*Ruppiaceae
*Rutaceae

*Salicaceae
Salvadoraceae
*Salviniaceae
Santalaceae
*Sapindaceae
Sapotaceae
*Schizaeaceae
Scrophulariaceae

Scytopetalaceae
Selaginellaceae
Selaginaceae — in Scrophulariaceae
*Simaroubaceae
*Smilacaceae
Sonneratiaceae
Sphenocleaceae
Strychnaceae — in Loganiaceae
*Surianaceae
Sterculiaceae

Taccaceae
Tamaricaceae
Tecophilaeaceae
Ternstroemiaceae — in Theaceae
Tetragoniaceae — in Aizoaceae
Theaceae
Thelypteridaceae
Thismiaceae — in Burmanniaceae
Thymelaeaceae
*Tiliaceae
Trapaceae
Tribulaceae — in Zygophyllaceae
*Triuridaceae
Turneraceae
Typhaceae

Uapacaceae — in Euphorbiaceae
Ulmaceae
*Umbelliferae
*Urticaceae

Vacciniaceae — in Ericaceae
Valerianaceae
Velloziaceae
*Verbenaceae
*Violaceae
*Viscaceae
*Vitaceae
*Vittariaceae

*Woodsiaceae

*Xyridaceae

*Zannichelliaceae
*Zingiberaceae
*Zosteraceae
*Zygophyllaceae

FORNTCOMING PARTS

Editorial adviser, National Museums of Kenya: Quentin Luke
Adviser on Linnaean types: C. Jarvis

Parts of this Flora, unless otherwise indicated, are obtainable from:
Royal Botanic Gardens, Kew, Richmond, Surrey TW9 3AB, England. www.kew.org or www.kewbooks.com

* only available through CRC Press at:
UK and Rest of World (except North and South America):
CRS Press/ITPS,
Cheriton House, North Way, Andover, Hants SP10 5BE.
e: uk.tandf@thomsonpublishingservices. co.uk

North and South America:
CRC Press,
2000NW Corporate Blvd, Boco Raton, FL 33431-9868,
USA.
e: orders@crcpress.com

Information on current prices can be found at www.kewbooks.com or www.tandf.co.uk/books/